Signaling System No. 7 (SS7) in Telecommunication Networks

Protocols, Architecture, Application, Implementation

Md. Munjure Mowla

Md. Mamunur Rashid

Md. Shohidul Islam

TABLE OF CONTENTS

Chapter 1: Introduction to Signaling

1.1 Signaling..1

    1.1.1 History..2

    1.1.2 Subscriber Signaling..3

        1.1.2.1 Address Signals..3

        1.1.2.2 Supervisory Signals...4

        1.1.2.3 Network Signaling...4

    1.1.3 Channel Associated Signaling (CAS).............................5

        1.1.3.1 Address Signals..6

        1.1.3.2 Supervisory Signals...6

        1.1.3.3 Limitations of CAS..6

    1.1.4 Common Channel Signaling (CCS)................................7

        1.1.4.1 Types of Common Channel Signaling..............8

            1.1.4.1.1 Signaling System No. 6 (SS6).............8

            1.1.4.1.2 Signaling System No. 7 (SS7).............9

        1.1.4.2 Common Channel Signaling Modes.................9

    1.1.5 Advantages of CCS over CAS..11

    1.1.6 Disadvantages of CCS in comparison to CAS...............11

1.2 Summary..12

Chapter 2: The Roles of SS7

2.1 Introduction..13

2.2 Signaling System No. 7-Based Services................................13

    2.2.1 Telephone-Marketing Numbers.....................................14

    2.2.2 Televoting..14

2.2.3 Single Directory Number…………………………………………..15

2.2.4 Enhanced 911…………………………………………………......15

2.2.5 Supplementary Services……………………………………………16

2.2.6 Custom local area signaling services (CLASS)…………………....17

2.2.7 Calling name (CNAM)……………………………………………18

2.2.8 Line Information Database (LIDB)………………………………..18

2.2.9 Local Number Portability (LNP)………………………………….18

2.2.10 2nd and 3rd Generation Cellular Networks……………………….19

       2.2.10.1 Short Message Service (SMS)……………………………19

       2.2.10.2 Enhanced Messaging Service (EMS)……………………..20

2.2.11 Virtual Private Network (VPN)…………………………………..20

2.2.12 Do-Not-Call Enforcement………………………………………..21

2.3 The Key to Convergence……………………………………………………..21

2.3.1 Internet Call Waiting and Internet Calling Name Services………...21

2.3.2 Click-to-Dial Applications…………………………………………22

2.3.3 Web-Browser-Based of Telecommunication Services…………….22

2.3.4 WLAN "Hotspot" Billing…………………………………………..23

2.3.5 Location-Based Games……………………………………………..23

2.4 Summary………………………………………………………………………23

Chapter 3: SS7 Network Architecture

3.1 Introduction……………………………………………………………………24

3.2 History of SS7…………………………………………………………………25

3.3 SS7 Network Architecture……………………………………………………..27

3.4 Signaling Links and Linksets………………………………………………….27

3.5 Routes and Routesets…………………………………………………………..28

3.6 Node Types………………………………………………………………………..29

3.7 Signal Transfer Point……………………………………………..………………29

       3.7.1 National STP…………………………………………………………30

       3.7.2 International STP……………………………………………………..30

       3.7.3 Gateway STP…………………………………………………………31

3.8 Service Switching Point…………………………………………………………..31

3.9 Service Control Point……………………………………………………………..32

3.10 Link Types……………………………………………………………………….32

       3.10.1 Access Links (A Links)……………………………………………..33

       3.10.2 Cross Links (C Links)………………………………………………33

       3.10.3 Bridge Links (B Links)……………………………………………..33

       3.10.4 Diagonal Links (D Links)…………………………………………..34

       3.10.5 Extended Links (E Links)…………………………………………..35

       3.10.6 Fully-Associated Links (F Links)…………………………………..36

3.11 Signaling Modes…………………………………………..……………………..37

3.12 Signaling Network Structure…………………………………………………….38

3.13 Summary…………………………………………………………………………41

Chapter 4: Protocols of SS7

4.1 SS7 Protocol Overview…………………………………………………………...42

4.2 SS7 Protocol Stack………………………………….....…………………………43

       4.2.1 SS7 Level 1: Physical Connection…………………………………...43

       4.2.2 SS7 Level 2: Data Link………………………………………………44

       4.2.3 SS7 Level 3: Network Level…………………………………..……..44

4.3 MTP……………………………………………………………………………….46

4.4 User and Application Parts………………………………………………………..48

4.4.1 TUP and ISUP………………………………………………………48

4.4.2 BISUP…………………………………….....…………………….49

4.5 SCCP………………………………………………………………………...49

4.6 TCAP………………………………………………………..……………...50

4.7 MAP……………………………………………………………..…………50

4.8 Summary………………………………………………..………………….51

Chapter 5: Message Transfer Part (MTP)

5.1 Introduction to MTP ……………………………………………………52

5.2 Different types of MTP………………………………......………………………52

    5.2.1 MTP 1 Functions……………………………………………….53

    5.2.2 MTP 2 Functions……………………………………………….53

        5.2.2.1 Fill-In Signal Units……………………………….....……….53

        5.2.2.2 Link Status Signal Units…………………………………….54

        5.2.2.3 Message Signal Units………………………………………55

        5.2.2.4 Field Descriptions………………………………..…………55

        5.2.2.5 Signal Unit Delimitation…………………………………….56

        5.2.2.6 Length Indicator……………………………………………57

        5.2.2.7 Signal Unit Alignment………………………..……………58

        5.2.2.8 Error Detection……………………………………………58

        5.2.2.9 Error Correction……………………………………………58

        5.2.2.10 Basic Error Correction……………………………………58

            5.2.2.10.1 Examples of Error Correction…..………………….…59

            5.2.2.10.2 Comparison with the Basic Error Correction Method..…60

        5.2.2.11 Signaling Link Initial Alignment……………..……………61

            5.2.2.11.1 Status Indications………………………………………61

    5.2.2.12 Signaling Link Error Monitoring……………………..…………63

      5.2.2.12.1 SUERM……………………………………………..……63

      5.2.2.12.2 AERM……………………………………………..……..64

    5.2.2.13 Flow Control………………………………………………..………64

  5.2.3 MTP 3 Functions………………………………………………..…………65

    5.2.3.1 Message Format………………………………………………..……65

    5.2.3.2 ITU-T Routing Label…………………………………..………..……65

    5.2.3.3 ANSI Routing Label……………………………………………..…..66

    5.2.3.4 Signaling Message Handling……………………………..…………66

    5.2.3.5 Message Load Sharing……………………………………………...68

    5.2.3.6 Comparing the IP and MTP3 Protocols…………..…………………69

    5.2.3.7 Signaling Network Management………………………..……………69

    5.2.3.8 Link Management………………………………………..…………..70

    5.2.3.9 Route Management…………………………………..………………71

    5.2.3.10 Traffic Management……………………………..…………………73

5.3 Summary………………………………………………………..………………74

Chapter 6: Signaling Connection Control Part (SCCP)

6.1 Introduction………………………………………………………..……………75

6.2 SCCP Architecture………………………………………………………………76

  6.2.1 SCCP Connectionless Control (SCLC)……………………..…………..77

  6.2.2 SCCP Connection-Oriented Control (SCOC)…………………………..78

  6.2.3 SCCP Routing Control (SCRC)…………………………………………79

    6.2.3.1 Subsystem Number (SSN) Routing…………………………………81

    6.2.3.2 Global Title Routing……………………………………..…………82

6.3 SCCP Message Transfer Services……………………………..………………83

  6.3.1 Connectionless Services………………………………………………..84

6.3.2 Connection-Oriented Services..85

6.4 SCCP Messages...86

    6.4.1 Message Structure...86

    6.4.2 Message Types..88

6.5 Summary..91

Chapter 7: ISDN User Part (ISUP)

7.1 Introduction...92

7.2 ISUP and the SS7 Protocol Stack..93

7.3 Basic ISUP Call Flow...94

7.4 ISUP Message Flow..95

7.5 ISUP Services..96

    7.5.1 Bearer Services... 96

    7.5.2 User Terminal Services.. 97

    7.5.3   Supplementary Services...97

        7.5.3.1 Call forwarding–unconditional (CFU)...............................98

        7.5.3.2 Call Forwarding-Busy (CFB)..98

        7.5.3.3 Call Forwarding–No Answer (CFNA)................................98

        7.5.3.4 Call Forwarding–Default (CFD)...98

        7.5.3.5 Calling Number Identification Presentation (CNIP).............99

        7.5.3.6 Calling Number Identification Restriction (CNIR)...............99

        7.5.3.7 Calling number identification restriction over (CNIR-Over)....99

7.6 ISUP Message..99

    7.6.1 ISUP Message Format...99

    7.6.2 Message types..101

    7.6.2.1 Initial Address Message (IAM)..102

    7.6.2.2 Address Complete Message (ACM)...102

    7.6.2.3 Answer Message (ANM)..103

    7.6.2.4 Release Message (REL)..103

    7.6.2.5 Release Complete Message (RLC)...104

  7.6.3 Subscriber Interface Message Tracing...104

7.7 ISUP and Local Number Portability..105

7.8 Interworking with ISDN..106

7.9 End-to-End Signaling..107

7.10 Summary..107

Chapter 8: Transaction Capabilities Application Part (TCAP)

8.1 Introduction..109

8.2 TCAP Structure 110

8.3 Role of TCAP in Call Control...110

8.4 TCAP within the SS7 Protocol Stack..111

8.5 Transaction and Component Sublayers..111

8.6 Message Layout..111

8.7 ITU Protocol Message Contents...112

  8.7.1 Unidirectional Message...113

  8.7.2 Begin Message...113

  8.7.3 End Message...114

  8.7.4 Continue Message...114

  8.7.5 Abort Message..115

8.8 Summary..116

Chapter 9: Mobile Application Part (MAP)

9.1 Introduction……………………………………………………………...117

9.2 MAP Services………………………………………………………….…119

    9.2.1 Mobility Services……………………………………………….119

    9.2.2 Operation and Maintenance…………………………………….120

    9.2.3 Call Handling………………………………………………..…..120

    9.2.4 Supplementary Services……………………………………...….120

    9.2.5 Short Message Service (SMS)………………………………..….121

9.3 MAP Operations ……………………………………………………….…122

9.4 MAP Messages……………………………………………………………122

    9.4.1 Format of MAP Messages……………………………………....122

9.5 Mobility Management and Call Processing……………………………….123

9.6 Summary……………………………………………………………….…126

Chapter 10: Practical Implementation of SS7 Signaling Link

10.1 Introduction…..……………………………………………………….....127

10.2 Explanation……………………………………………………………....127

10.3 Summary ……………………………………………………………......131

Chapter 11: Conclusion and Future Recommendations

11.1 Conclusion ………………………………………………………….…...131

11.2 Future Recommendations………………………………………………..132

REFERENCES………………………………………………………......133

LIST OF TABLES

Table No.	Title	Page No.
Table 3-1:	CCITT/ITU-T SS7 Timeline	25
Table 5-1:	Values in the Status Field	55
Table 5-2:	Field Descriptions	56
Table 5-3:	Comparison of IP and MTP3 Packet Fields	69
Table 6-1:	SCCP message type and code	88
Table 8-1:	Unidirectional Message Fields	113
Table 8-2:	Begin Message Fields	113
Table 8-3:	End Message Fields	114
Table 8-4:	Continue Message Fields	114
Table 8-5:	Abort Message Fields	115
Table 10-1:	MTP Point Code Information	128
Table 10-2:	MTP Link Table Information	128
Table 10-3:	MTP Route Table Information	129

LIST OF FIGURES

Figure No.	Title	Page No.
Figure 1-1:	Dial Pulse Address Signals	3
Figure 1-2:	Subscriber and Network Signaling	5
Figure 1-3:	Associated Mode	10
Figure 1-4:	Quasi-Associated Mode	11
Figure 3-1:	How Many Pages the ITU C7 Specifications Covered Based on Year	26
Figure 3-2:	Four Links in a Linkset Between SPs	28
Figure 3-3:	Route from SP A to SP C	28
Figure 3-4:	Route set from SP A to SP C	29
Figure 3-5:	SS7 Node Types	29
Figure 3-6:	A Links	33
Figure 3-7:	C Links	33
Figure 3-8:	B Links	34
Figure 3-9:	D Links	34
Figure 3-10:	Existence of an STP Backbone and STP Hierarchy	35
Figure 3-11:	E Links	36
Figure 3-12:	F Links	36
Figure 3-13:	SS7 Network Segment	37
Figure 3-14:	SS7 Signaling Modes	38
Figure 3-15:	Typical Example of North American SSP Interconnections	39
Figure 3-16:	Typical Example of SSP Interconnections in Most Areas Outside North America	39
Figure 3-17:	Example of Direct and Indirect SSP Interconnections to STPs	40
Figure 4-1:	Introductory SS7 Protocol Stack	42
Figure 4-2:	SS7 Protocol Stack	44
Figure 4-3:	A Single MTP3 Controls Many MTP2s, Each of Which Is Connected to a Single MTP1	48
Figure 4-4:	SS7 Level 4 Protocols, User and Application Parts	48
Figure 5-1:	Position of the MTP in SS7	52

Figure 5-2:	FISU Format	54
Figure 5-3:	LSSU Format	54
Figure 5-4:	MSU Format	55
Figure 5-5:	Zero Bit Insertion and Deletion	57
Figure 5-6:	Principles of Basic Error Correction	60
Figure 5-7:	Procedure for Signaling Link Alignment	63
Figure 5-8:	Flow Control Using Status Indication SIB	64
Figure 5-9:	ITU-T Routing Label	65
Figure 5-10:	ANSI Routing Label	66
Figure 5-11:	Signaling Message Handling	67
Figure 5-12:	How Loss of Linkset Affects Routes	71
Figure 6-1:	SS7 Stack with the Network Service Part (NSP) Highlighted	75
Figure 6-2:	The SCCP Architecture	77
Figure 6-3:	The Transfer of Connectionless Messages from One SCCP User to Another	77
Figure 6-4:	The Transfer of Connection-oriented Messages from One SCCP User to Another Using a Temporary Connection	79
Figure 6-5:	Showing MTP Point Code Routing	81
Figure 6-6:	An SSN and DPC Are Required for the Final Delivery of an SCCP Message	81
Figure 6-7:	Example of GTT	83
Figure 6-8:	Connectionless transfer of signaling messages	84
Figure 6-9:	Connection-oriented transfer with a middle node	85
Figure 6-10:	Structure of SCCP messages	86
Figure 6-11:	An Example of a Connection Request (CR) Message Structure	88
Figure 7-1:	ISUP at Level 4 of the SS7 Stack	93
Figure 7-2:	Simple ISUP Message Flow	95
Figure 7-3:	ISUP Message Format	100
Figure 7-4:	IAM Message Format	102
Figure 7-5:	ACM Message Format	103
Figure 7-6:	REL Message Format	103
Figure 7-7:	Interface Message Tracing in Same Operator	104
Figure 7-8:	Interface Message Tracing in Different Operator	104

Figure 7-9:	ISUP-ISDN Interworking..106
Figure 7-10:	ISUP End-to-End Signaling..107
Figure 8-1:	Position of TCAP in the SS7 network..................................109
Figure 8-2:	The structure of TCAP 110
Figure 8-3:	TCAP within the SS7 Stack..111
Figure 8-4:	TCAP Message Structure..112
Figure 9-1:	Mobile network architecture...118
Figure 9-2:	Structural relation between MAP and MTP messages.........122
Figure 9-3:	Placing an MTC..125
Figure 10-1:	Different SPs location showing direct and STP connection of a telecom core network..127

LIST OF ACRONYMS AND ABBREVIATIONS

AC - Authentication Center

ACM - Address Complete Message

ANSI - American National Standard Institute

ASCII - American Standard Code for Information Interchange

BCD- Binary-Coded Data

BLA - Block Acknowledge signal

BLO - Blocking signal

BS - Base Station

BSAP - Base Station Application Part

BSC - Base Station Controller

BSM - Backward Set-up Message

BSMAP - BS Management Application Part

BSN - Backward Sequence Number

BSSMAP - Base Station Subsystem Management Application Part

CC -Connection Confirm signal

CCL -Calling Party Clear signal

CCM - Circuit Supervision Message

CCR - Continuity Check Request

CDMA - Code Division Multiple Access

CGB - Circuit Group Blocking

CGBA- Circuit Group Blocking ACK Technical Manual – Signaling System

CGU -Circuit Group Unblocking

CGUA -Circuit Group Unblocking ACK

CIC -Circuit Identification Code

CIE -Content of Information Element

CK - Check bit

CLIP -Calling Line Identification Presentation

CM -Connection Management

CNM -Circuit Network Management message group

CO -Connection Oriented

COT -Continuity Signal

CPG -Call Progress message

CR -Connection Request message

CREF -Connection Refused message

CSL -Component Sublayer

CSM -Call Supervision Message

CUG - Closed User Group

DL- Digital Line Signaling

DLCI -Data Link Connection Identifier

DPC -Destination Point Code

DSS1 -Digital Subscriber Signaling No.1

DTAP- Direct Transfer Application Part

DTMF - Dual-Tone Multi-Frequency

DUP - Data User Part

ESN -Electronic Serial Number

FAA -Facility Accepted message

FAM -Forward Address message

FAR -Facility Request message

FIB -Forward Indicator Bit

FISU -Fill-In Signal Unit

FOT -Forward Transfer signal

FRJ -Facility Reject message

FSM -Forward Setup Message

FSN- Frame Switching Network Board

GT -Global Title

HDLC -High-speed Digital Link Channel

HLR -Home Location Register

HSTP -High-level Signaling Transfer Point

IAI -Initial Address Message with Information

IAM- Initial Address Message

ID -Identification

IE -Information Element

IEI -Information Element Identification

IMSI -International Mobile Station Identity

INAP -Intelligent Application Part

INR -Information Request

IP -Intelligent Peripherals

IPM -Intelligent Peripherals Model

ISC -International Switching Center

ISDN- Integrated Services Digital Network

ISUP -ISDN User Part

IT -Inactivity Test

ITU-T -International Telecommunication Union - Telecommunication

IWF -Interworking Function

LAN -Local Area Network

LAPD -Link Access Protocol on the D-channel

LI -Length Indicator

LSSU -Link Status Signal Unit

LSTP -Low-Level Signaling Transfer Point

MAP -Mobile Application Part

MC -Message Center

MCI -Malicious Call Identification

MFC -Multiple Frequency Control

MIN -Mobile Identification Number

MML -Man-Machine Language

MODEM -Modulator-Demodulator

MSC -Mobile Switching Center

MSU -Message Signal Unit

MTP -Message Transfer Part

NAT -National Network

NCB -National Call Supervision message

NSB -National Successful Backward Setup message

NSP -Network Service Part

NSS -Network Subsystem

NT2 -Network Termination 2

OMAP -Operation and Maintenance Application Part

OPC -Originating Point Code

OSI -Open System Interconnection

PABX -Private Automatic Branch Exchange

PCM -Pulse Code Modulation

PIN -Personal Identification Number

PLMN -Public Land Mobile Network

PPC -Pre-Paid Charging

PRI -Primary Rate Interface

PSTN -Public Switched Telephone Network

RAND -Random number

RandC -Random Challenge

REL -Release

RES -Resume

RSC -Reset Circuit

SAM -Subsequent Address Message

SAP -Service Access Point

SCCP -Signaling Connection and Control Part

SCP -Service Control Point

SDU -Selection/Distribution Unit

SI -Service Indicator

SIF -Signaling Information Field

SIO -Service Information Octet

SLC -Signaling Link Code

SLS -Signaling Link Selection

SP -Signaling Point

SPC -Signaling Point Code

SS7 -Signaling System No.7

SSF -Sub-Service Field

SSN -Subsystem Number

STP -Signaling Transfer Point

SU -Single Unit

SUS -Suspend

TC -Transaction Capability

TCAP -Transaction Capabilities Application Part

TE -Terminal Equipment

TLDN -Temporary Mobile Directory Number

TOD - Time of Day

TRIG -Trigger

TSL -Transaction Sublayer

TUP -Telephone User Part

UBA -Unblocking Acknowledgment signal

UBL -Unblocking Signal

UD -User Data

UDT -Unit Data

UDTS -Unit Data Service

UP -User Part

VLR -Visitor Location Register

VPN -Virtual Private Network

XUDT- Enhanced Unit Data

XUDTS - Enhanced Unit Data Service

Chapter 1

Introduction to Signaling

1.1 Signaling

The ITU-T defines signaling as, "The exchange of information (other than by speech) specifically concerned with the establishment, release and other control of calls, and network management, in automatic telecommunications operation." [1]

Signaling System No. 7 is known commonly in North America it is often referred to as CCSS7, an acronym for Common Channel Signaling System 7. In some European countries, specifically the UK, it is sometimes called C7 (CCITT number 7) and is also known as number 7 and CCIS7 (Common Channel Interoffice Signaling 7).

The purpose of network signaling is to set up a circuit between the calling and called parties so that user traffic (voice, fax, and analog dial-up modem, for example) can be transported bi-directionally. When a circuit is reserved between both parties, the destination local switch places a ringing signal to alert the called party about the incoming call. This signal is classified as subscriber signaling because it travels between a switch (the called party's local switch) and a subscriber (the called party). A ringing indication tone is sent to the calling party telephone to signal that the telephone is ringing. If the called party wishes to engage the call, the subscriber lifts the handset into the off-hook condition. This moves the call from the set-up phase to the call phase.

The main function of signaling is still that of circuit supervision: setting up and clearing down circuits (that is, trunks). Traditionally, once a circuit was set up, no other signaling was performed apart from releasing the call; therefore, all calls were simple, basic telephone service calls. However, modern telephone networks can perform signaling while a call is in progress, especially for supplementary services—for example, to introduce another called party into the call, or to signal the arrival of another incoming call (call waiting) to one of the parties. In fact, since the 1980s, signaling can take place even when there is not a call in place. This is known as non-circuit related signaling and is simply used to transfer data between networks nodes. It is primarily used for query and response with telecommunications databases to support cellular networks, intelligent networks, and supplementary services. For example, in Public Land Mobile Networks (PLMNs), the visitor location register (VLR) that is in charge of the area into which the subscriber has roamed updates the home location register (HLR) of the subscriber's location. PLMNs make much use of non-circuit-related signaling, particularly to keep track of roaming subscribers.

This is a high-level view what signaling networks really do. They enable telephone switches (now also packet switches) to communicate with one another and share information needed to process any type of call autonomously. SS7 was originally designed for analog telephone networks but this has established itself to ever changing world of telecommunications. [2]

1.1.1 History

Common Channel Signaling protocols have been developed by major telephone companies and the ITU-T since 1975 and the first international Common Channel Signaling protocol was defined by the ITU-T as Signaling System No. 6 (SS6) in 1977. Signaling System No. 7 was defined as an international standard by ITU-T in its 1980. SS7 was designed to replace SS6, which had a restricted 28-bit signal unit that was both limited in function and not amenable to digital systems. SS7 has substantially replaced SS6, Signaling System No. 5 (SS5), R1 and R2, with the exception that R1 and R2 variants are still used in numerous nations.[3]

SS5 and earlier systems used in-band signaling, in which the call-setup information was sent by playing special multi-frequency tones into the telephone lines, known as bearer channels in the parlance of the telecom industry. This led to security problems with blue boxes. SS6 and SS7 implement out-of-band signaling protocols, carried in a separate signaling channel, explicitly keep the end-user's audio path—the so-called speech path—separate from the signaling phase to eliminate the possibility that end users may introduce tones that would be mistaken for those used for signaling. SS6 and SS7 are referred to as so-called Common Channel Interoffice Signaling Systems (CCIS) or Common Channel Signaling (CCS) due to their hard separation of signaling and bearer channels. This required a separate channel dedicated solely to signaling, but the greater speed of signaling decreased the holding time of the bearer channels, and the number of available channels was rapidly increasing anyway at the time SS7 was implemented.

The common channel signaling paradigm was translated to IP via the SIGTRAN protocols as defined by the IETF. While running on a transport based upon IP, the SIGTRAN protocols are not an SS7 variant, but simply transport existing national and international variants of SS7.[4]

1.1.2 Subscriber Signaling

Subscriber signaling takes place on the line between the subscribers and their local switch. Most subscribers are connected to their local switch by analog subscriber lines as opposed to a digital connection provided by an Integrated Services Digital Network (ISDN). As a result, subscriber signaling has evolved less rapidly than network signaling.

Subscriber signals can be broken down into the following four categories:

1. Address Signals
2. Supervisory Signals
3. Tones and Announcements
4. Ringing

1.1.2.1 Address Signals

Address signals represent the called party number's dialed digits. Address signaling occurs when the telephone is off-hook. For analog lines, address signaling is either conveyed by the dial pulse or Dual-Tone Multiple Frequency (DTMF) methods. Local switches can typically handle both types of address signaling, but the vast majority of subscribers now use Dual-Tone Multi Frequency (DTMF), also known as touch-tone.

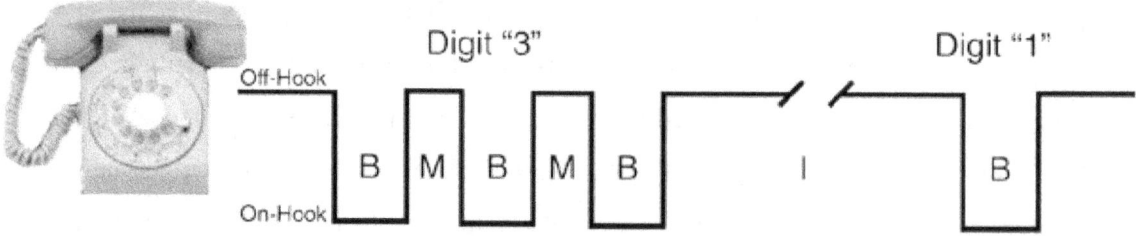

Figure 1-1: Dial Pulse Address Signals

1.1.2.2 Supervisory Signals

A telephone has two possible supervision states: on-hook or off-hook. On-hook is the condition in which the telephone is not in use, which is signaled when the telephone handset depresses the cradle switch. The term on-hook comes from the days when the receiver part of the telephone rested on a hook. The telephone enters the off-hook condition when the handset is lifted from its cradle; thereby releasing the cradle switch and signaling to the exchange that the subscriber wishes to place an outgoing call.

Residential systems worldwide use a change in electrical conditions, known as loop start signaling, to indicate supervision signals. The local switch provides a nominal – 48 V direct current (DC) battery, which has the potential to flow through the subscriber line (between the local switch and the subscriber). When a telephone is off-hook, DC can flow in the subscriber line; when a telephone is on-hook a capacitor blocks the DC. The presence or absence of direct current in the subscriber's local switch line determines the telephone's supervision state. Loop start systems are adequate for residential use, but a problem known as glare makes loop start unacceptable in typical business applications in which private exchanges (PBXs) are used. PBXs use a system known as ground start signaling, particularly in North America.

Ground start systems combat glare by allowing the network to indicate off-hook (seizure) for incoming calls, regardless of the ringing signal. This reduces the probability of simultaneous seizure, or glare, from both ends. Ground start requires both ground and current detectors in customer premise equipment (CPE).

1.1.2.3 Network Signaling

As previously described, network signaling takes place between nodes in the core network. This is generally from the local switch, through the core network, and to the destination local switch—in other words, between the calling and the called party switch. Figure 1-2 shows where subscriber and network signaling occur in the PSTN.

Figure 1-2: Subscriber and Network Signaling

Network signaling was previously implemented using Channel Associated Signaling (CAS) techniques and systems. However, for the past two decades, it has been replaced with Common Channel Signaling (CCS) systems. Apart from a rare trace of Signaling System No. 6 (SS6) signaling, System No. 7 (SS7) is almost the exclusive CSS system; thus, CCS can almost be taken to refer exclusively to the use of SS7. [2]

1.1.3 Channel Associated Signaling

Channel Associated Signaling (CAS), also known as per-trunk signaling (PTS), is a form of digital communication signaling. As with most telecommunication signaling methods, it uses routing information to direct the payload of voice or data to its destination. With CAS signaling, this routing information is encoded and transmitted in the same channel as the payload itself. This information can be transmitted in the same band (in-band signaling) or a separate band (out-of-band signaling) to the payload. [5]

CAS can be implemented using the following related systems:

- Bell Systems MF, R2, R1, and C5.
- Single-frequency (SF) in-band and out-of-band signaling
- Robbed bit signaling

1.1.3.1 Address Signals

Multifrequency systems, such as the Bell System MF, R2, R1, and C5, are all types of address signals used by CAS.

The CAS system can be used on either analog Frequency Division Multiplexed (FDM) or digital Time Division Multiplexed (TDM) trunks. MF is used to signal the address digits between the switches.

1.1.3.2 Supervisory Signals

Single frequency systems, robbed bit signaling, and digital signaling are all types of supervisory signals used by CAS.

Single Frequency (SF) was used for supervisory signaling in analog CAS-based systems. North America used a frequency of 2600 Hz (1600 Hz was previously used), and Great Britain used 2280 Hz (as defined in British Telecom's SSAC15 signaling specification). When in an on-hook state, the tone is present; when in an off-hook state, the tone is dropped.

1.1.3.3 Limitations of CAS

CAS has a number of limitations, including:

- Susceptibility to fraud
- Limited signaling states
- Poor resource usage/allocation

The following sections discuss these limitations in more detail.

Susceptibility to Fraud

CAS employing in-band supervisory signaling is extremely susceptible to fraud because the subscriber can generate these signals by simply using a tone generator down a handset mouthpiece. This type of device is known as a blue box; from the beginning of the 1970s, it could be purchased as a small, handheld keypad. Blue box software was available for the personal computer by the beginning of the 1980s.

Limited Signaling Information

CAS is limited by the amount of information that can be signaled using the voice channel. Because only a small portion of the voice band is used for signaling, often CAS cannot meet the requirements of today's modern networks, which require much higher bandwidth signaling.

Inefficient Use of Resources

CAS systems are inefficient because they require either continuous signaling or, in the case of digital CAS, at regular intervals even without new signals.

In addition, there is contention between voice and signaling with in-band CAS. As a result, signaling is limited to call set-up and release phases only. This means that signaling cannot take place during the call connection phase, severely imposing technological limits on the system's complexity and usefulness. [2]

1.1.4 Common Channel Signaling (CCS)

In telephony, Common Channel Signaling (CCS), in the US also Common Channel Interoffice Signaling (CCIS), is the transmission of signaling information (control information) on a separate channel from the data, and, more specifically, where that signaling channel controls multiple data channels.[3]

For example, in the public switched telephone network (PSTN) one channel of a communications link is typically used for the sole purpose of carrying signaling for establishment and tear down of telephone calls. The remaining channels are used entirely for the transmission of voice data. In most cases, a single 64kbit/s channel is sufficient to handle the call setup and call clear-down traffic for numerous voice and data channels.

The logical alternative to CCS is Channel Associated Signaling (CAS), in which each bearer channel has a signaling channel dedicated to it.

CCS offers the following advantages over CAS, in the context of the PSTN:

- Faster call setup.

- No falsing interference between signaling tones by network and speech frequencies.
- Greater trunking efficiency due to the quicker set up and clear down, thereby reducing traffic on the network.
- No security issues related to the use of in-band signaling with CAS.
- CCS allows the transfer of additional information along with the signaling traffic providing features such as caller ID.
- The most common CCS signaling methods in use today are Integrated Services Digital Network (ISDN) and Signaling System 7 (SS7).

ISDN signaling is used primarily on trunks connecting end-user private branch exchange (PBX) systems to a central office. SS7 is primarily used within the PSTN. The two signaling methods are very similar since they share a common heritage and in some cases, the same signaling messages are transmitted in both ISDN and SS7.

CCS is distinct from in-band or out-of-band signaling, which are to the data band what CCS and CAS are to the channel.

1.1.4.1 Types of Common Channel Signaling

Two types of CCS were available namely-
- Signaling System No. 6 (SS6)
- Signaling System No. 7 (SS7)

1.1.4.1.1 Signaling System No. 6 (SS6)

Signaling System No. 6 (SS6) was introduced in the 1970s as an early common channel signaling method for telephone trunks between International Switching Centers (ISCs). It had applications as there was a need to experience common channel working on a fast digital basis. The SS6 signaling channel was typically a 2.4 kbit/s data link.

SS6 was a method developed & implemented by administrations in a limited number of countries for use in correspondent International Switching of analogue telephone calls. The next development was correspondent and non-correspondent switching of telephone calls by new operators in the counties where more sophisticated methods were required.

1.1.4.1.2 Signaling System No. 7 (SS7)

Signaling System No. 7 (SS7) is a set of telephony signaling protocols which are used to set up most of the world's public switched telephone network telephone calls. The main purpose is to set up and tear down telephone calls. Other uses include number translation, prepaid billing mechanisms, short message service (SMS), and a variety of other mass market services.

SS7 signaling is termed Common Channel Signaling (CCS) in that the path and facility used by the signaling is separate and distinct from the telecommunications channels that will ultimately carry the telephone conversation. With CCS, it becomes possible to exchange signaling without first seizing a facility, leading to significant savings and performance increases in both signaling and facility usage.

1.1.4.2 Common Channel Signaling Modes

A signaling mode refers to the relationship between the traffic and the signaling path. Because CCS does not employ a fixed, deterministic relationship between the traffic circuits and the signaling, there is a great deal of scope for the two to have differing relationships to each other. These differing relationships are known as **signaling modes**.

There are three types of CCS signaling modes:

- Associated
- Quasi-associated
- Non-associated

SS7 runs in associated or quasi-associated mode, but not in non-associated mode. Associated and quasi-associated signaling modes ensure sequential delivery, while non-associated does not. SS7 does not run in non-associated mode because it does not have procedures for reordering out-of-sequence messages.

Associated Signaling

In associated mode, both the signaling and the corresponding user traffic take the same route through the network. Networks that employ only associated mode are easier to design and maintain; however, they are less economic, except in small-sized networks. Associated mode requires every network switch to have signaling links to every other interconnected switch (this is known as a fully meshed network design). Usually a minimum of two signaling links are employed for redundancy, even though the switched traffic between two interconnected switches might not justify such expensive provisioning. Associated signaling mode is the common means of implementation outside of North America. Figure 1-3 illustrates the associated concept.

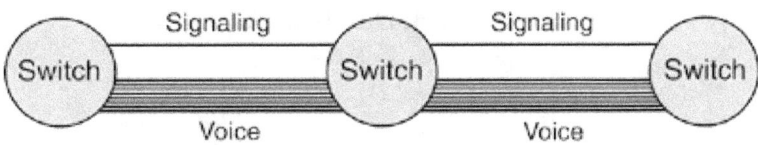

Figure 1-3: Associated Mode

Quasi-Associated Signaling

In quasi-associated mode, signaling follows a different route than the switched traffic to which it refers, requiring the signaling to traverse at least one intermediate node. Quasi-associated networks tend to make better use of the signaling links; however, it also tends to create a more complex network in which failures have more potential to be catastrophic.

Quasi-associated signaling can be the most economical way of signaling for lightly loaded routes because it avoids the need for direct links. The signaling is routed through one or more intermediate nodes. Signaling packets arrive in sequence using quasi-associated signaling because the path is fixed for a given call (or database transaction) at the start of a call (or transaction). Figure 1-4 shows the quasi-associated signaling mode, which is the common means of implementation within North America.

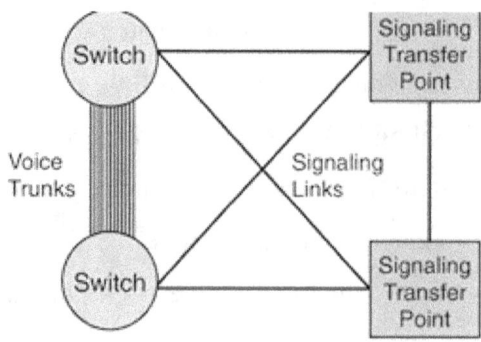

Figure 1-4: Quasi-Associated Mode

Non-Associated Signaling

Because the path is not fixed at a given point in time in non-associated mode, the signaling has many possible routes through the network for a given call or transaction. Therefore, the packets might arrive out of sequence because different routes might have been traversed.

SS7 does not run in non-associated mode because no procedures exist for reordering out-of-sequence messages. Associated and quasi-associated signaling modes assure sequential delivery, while non-associated signaling does not. Quasi-associated mode is a limited case of non-associated mode, in which the relative path is fixed.

1.1.5 Advantages of CCS over CAS

- Much faster call set-up time
- Greater flexibility
- Capacity to evolve
- More cost effective than CAS
- Greater call control

1.1.6 Disadvantages of CCS in comparison to CAS

- CCS links can be a single point of failure—a single link can control thousands of voice circuits, so if a link fails and no alternative routes are found, thousands of calls could be lost.

- There is no inherent testing of speech path by call set-up signaling, so elaborate Continuity Test procedures are required.[2]

1.2 Summary

This chapter deals with the definition of signaling and comparison among various signaling system. The advantage and limitation of different signaling system is also discussed in this chapter.

Chapter 2

The Roles of SS7

2.1 Introduction

SS7/C7 is the protocol suite that is employed globally, across telecommunications networks, to provide signaling; it is also a private, "behind the scenes," packet-switched network, as well as a service platform. Being a signaling protocol, it provides the mechanisms to allow the telecommunication network elements to exchange control information.

Over the past quarter of a century, SS7 has undergone a number of revisions and has been continually enhanced to support services that are taken for granted on a daily basis.

SS7/C7 is the key enabler of the public switched telephone network (PSTN), the integrated services digital network (ISDN), intelligent networks (INs), and public land mobile networks (PLMNs).[2]

2.2 Signaling System No. 7-Based Services

In addition to setting up and releasing calls, SS7/C7 is the workhorse behind a number of telecommunication services, namely-

- Telephone-marketing numbers such as toll-free and freephone
- Televoting (mass calling)
- Single Directory Number
- Enhanced 911 (E911)—used in the United States
- Supplementary services
- Custom local area signaling services (CLASS)
- Calling name (CNAM)
- Line information database (LIDB)
- Local number portability (LNP)
- Cellular network mobility management and roaming
  - Short Message Service (SMS)

- Enhanced Messaging Service (EMS)—Ringtone, logo, and cellular game delivery

- Local exchange carrier (LEC) provisioned private virtual networks (PVNs)
- Do-not-call enforcement

These telecommunications services are describe shortly below.

2.2.1 Telephone-Marketing Numbers

The most commonly used telephone-marketing numbers are toll-free calling numbers (800 calling), known as freephone (0800) in the United Kingdom. Because the call is free for the caller, these numbers can be used to win more business by increasing customer response. [2]

The advantage of using marketing numbers is that it gives one a national rather than a local presence - making ones business look bigger than it is. The same marketing number can follow one to anywhere in the UK - one do not need to change ones contact number even if one move ones premises.

The marketing numbers those are available on the market now including **0800/0870/0871/0905/0906** numbers. [6]

2.2.2 Televoting

Televoting, telephone voting or phone voting is a method of decision making and opinion polling conducted by telephone. Televoting can also extend to voting by SMS text message via a mobile cell phone.

Televoting or televote is the term used to describe the telephone voting in which broadcasters provide the audience with different telephone numbers associated with contestants participating and the outcome is decided by the number of calls to each line. Music contests such as the Eurovision Song Contest, as well as for World Idol, American Idol and similar use this method.

Televoting is a more cost-effective method of democratic deliberation than many alternatives such as deliberative polling, as it does not require the participants to meet in person.

Common to other deliberative democratic techniques, it also tends to produce more reasoned decisions than "raw" opinion polling, because participants are exposed to various perspectives other than their own in the briefing materials that they receive. [7]

2.2.3 Single Directory Number

Another service that uses SS7/C7 and has been deployed in recent years is the single directory number, which allows a company with multiple offices or store locations to have a single directory number. After analyzing the calling party's number, the switch directs the call to a local branch or store. [2]

2.2.4 Enhanced 911

Enhanced 911, E-911 or E911 in North America is one example of the modern evolution of telecommunications based system meant as an easy way to link people experiencing an emergency with the public resources that can help. The dial-three-digits concept was first originated in the United Kingdom in 1937. It has spread to continents and countries across the globe. Today other easy dial codes including 112 that were adopted by the European Union in 1991 to provide free-of-charge calling to those who need help during emergencies. The Emergency telephone number article contains comprehensive information regarding other emergency dialing codes for countries outside North America. [8]

E911 might also provide other significant location information, such as the location of the nearest fire hydrant, and potentially the caller's key medical details. The Federal Communications Commission (FCC) also has a cellular 911 program in progress; in addition to providing the caller's telephone number, this program sends the geographical location of the antenna to which the caller is connected. Enhancement proposals are already underway to obtain more precise location information.

A 911 address contains a uniform number, the street name, direction (if any) and the city. The address number is assigned usually by the grid of the existing community

2.2.5 Supplementary Services

Supplementary services provide the subscribers with more than plain old telephony service (POTS), without requiring them to change their telephone handsets or access technology. GSM Supplementary Services are primarily voice services and are usually the first to be offered by a network operator after providing the basic GSM Service. It offers people many benefits to suit needs for them. Well-known supplementary services include

- ✓ ***Call Conference (CF)***
- ✓ ***Call Barring (CB)***
- ✓ Call forwarding–unconditional (CFU)
- ✓ Call forwarding–busy (CFB)
- ✓ Call forwarding–no answer (CFNA)
- ✓ Call forwarding–default (CFD)
- ✓ Calling number identification presentation (CNIP)
- ✓ Calling number identification restriction (CNIR)
- ✓ Calling number identification restriction over (CNIR-Over)
- ✓ Call waiting (CW)
- ✓ Call transfer (CT)

Call Waiting (CW)

Call Waiting Service allows you to find out if someone's calling you, even while you're busy on another conversation. The caller will hear the phone ring instead of an engage tone. The incoming call can be attended by you and the first caller can be kept on hold. You can also swap between two incoming calls easily. [9]

Call Forwarding Unconditional (CFU)

Call forwarding Unconditional Service automatically diverts all incoming calls to a predefined number. The number can be a Standard Telephone, Mobile or Voice Mail number. [9]

Call conference (CC)

Call conference service allows you to make calls to more than one number at a time and have a multi-party conversation up to five people at a time, and each person you call can add up to 5 more participants. [9]

Call Barring (CB)

If one has a GSM postpaid line he can benefit from the following call barring options:

- ✓ Bar all outgoing international calls
- ✓ Bar all Incoming calls
- ✓ Bar all incoming calls while roaming abroad.

Recently, supplementary services have been helpful in increasing operators' revenues since revenues against call minutes have been on the decline. Usually the subscriber must pay a fixed monthly or quarterly fee for a supplementary service. [9]

2.2.6 Custom local area signaling services (CLASS)

Custom local area signaling services (CLASS) are an extension of supplementary services that employ the use of SS7 signaling between exchanges within a local geographical area. Information provided over SS7 links, such as the calling party number or the state of a subscriber line, enable more advanced services to be offered by service providers. A few examples of CLASS services include:

- ✓ Call block— Stops pre-specified calling party numbers from calling.
- ✓ Distinctive ringing— provides a distinct ringing signal when an incoming call originates from a number on a predefined list. This feature is particularly beneficial to households with teenagers.

- ✓ Priority ringing— provides a distinct ring when a call originates from a pre-specified numbers. If the called subscriber is busy and has call waiting, the subscriber receives a special tone indicating that a number on the priority list is calling.
- ✓ Call completion to busy subscriber (CCBS) — if a subscriber who has CCBS calls a party who is engaged in another call, the subscriber can activate CCBS with a single key or sequence. When activated, CCBS causes the calling party's phone to ring when the called party becomes available; when the calling party answers, the called party's phone automatically rings again. This feature saves the calling party from continuously attempting to place a call to a party is still unavailable. [2]

2.2.7 Calling name (CNAM)

Calling name (CNAM) is an increasingly popular database-driven service. With this service, the called party receives the name of the person calling in addition to their number. The called party must have a compatible display box or telephone handset to use this service.

The CNAM information is typically stored in regional telecommunications databases. SS7/C7 queries the database for the name based on the number and delivers the information to the called party's local switch. CNAM is a terminating user feature allowing a customer premises equipment (CPE) connected to a switching system to receive a calling party's name during the first silent interval.

2.2.8 Line Information Database (LIDB)

A Database application program that resides on a service control point (SCP) in the SS7 telephone signaling network and provides validation information for use in alternate billing services, such as telephone calling cards. Alternate billing services allow subscribers to bill calls to an account that is not necessarily associated with the originating line. For example, it can be used to validate a subscriber's calling card number that is stored in the LIDB, designating this as the means of payment. LIDB data base contains up-to-date records of all working lines, including directory listing name, description of the type of dialing capability subscribed (rotary dial vs. touch-tone), calling card numbers, and other data required for validation services.[10]

SS7/C7 is responsible for the real-time database query/response that is necessary to validate the calling card before progressing to the call setup phase.

2.2.9 Local Number Portability (LNP)

Local Number Portability (LNP) is the ability of a telephone customer to retain their local phone number if they switch to another local telephone service provider. There are three phases of number portability:

- ✓ Service Provider Portability
- ✓ Service Portability
- ✓ Location Portability

Generally porting a number involves

- ✓ Filling out a form
- ✓ Copy of an old bill
- ✓ Call Logs of old Number

LNP is primarily aimed at stimulating competition among providers by removing the personal inconvenience of changing phone numbers when changing service providers. For example, many businesses and individuals spend relatively large sums of money to print their phone numbers on business cards, letterheads, and other correspondence items.

2.2.10 2nd and 3rd Generation Cellular Networks

Cellular networks use SS7/C7 for the same reasons they use fixed line networks, but they place much higher signaling demands on the network because of subscriber mobility. All cellular networks, from 2G (GSM, ANSI-41, and even PDC, which is used in Japan) to 3G (UMTS and cdma2000), use SS7/C7 for call delivery, supplementary services, roaming, mobility management, prepaid, and subscriber authentication.

2.2.10.1 Short Message Service (SMS)

Short message service is a mechanism of delivery of short messages over the mobile networks. It is a store and forward way of transmitting messages to and from mobiles. The message (text only) from the sending mobile is stored in a central short message center (SMS) which then forwards it to the destination mobile. [11]

Cellular operators usually use SMS to alert the subscribers that they have voice mail, or to educate them on how to use network services when they have roamed onto another network. Third party companies offer the additional delivery services of sending SMS-to-fax, fax-to-SMS, SMS-to-e-mail, e-mail-to-SMS, SMS-to-web, web-to-SMS, and SMS notifications of the arrival of new e-mail.

Some of the common applications of SMS are:

- ✓ Exchanging small messages
- ✓ Many operators offer e-mail service over SMS.
- ✓ Information services like news, weather, entertainment and stock prices etc.
- ✓ SMS can be used by the network operators to provide services like balance enquiry in case of prepaid cards using SMS.
- ✓ It provides an alternative to alphanumeric paging services

2.2.10.2 Enhanced Messaging Service (EMS)

EMS (Enhanced Messaging Service) adds powerful new functionality to the popular text-based SMS (Short Messaging Service). Using EMS, mobile phone users can add life to their Text Messages in the form of images, melodies, and animations. [12]

Alcatel, Ericsson, Motorola and Siemens have committed to supporting EMS and bringing exciting images and sound to text messaging via EMS. The EMS standard was defined by 3GPP (3rd Generation Partnership Project), the same standardization body from which the worldwide success of GSM and the standardization of SMS (Short Messaging Services) originated.

Since EMS is simply an enhanced SMS service, it uses the SS7/C7 network in the same way; the SS7/C7 network carries it, and it uses SS7/C7 for the required signaling procedures.

2.2.11 Virtual Private Network (VPN)

A virtual private network (VPN) is a secure way of connecting to a private Local Area Network at a remote location, using the Internet or any unsecure public network to transport the network data packets privately, using encryption. The VPN uses authentication to deny access to unauthorized users, and encryption to prevent unauthorized users from reading the private network packets. The VPN can be used to send any kind of network traffic securely, including voice, video or data. [13]

SS7/C7 signaling (and a connected database) monitors the "private customer" line. The customer has all the features of a leased-line service as well as additional features, such as the ability to request extra services ad hoc and to tailor the service to choose the cheapest inter-exchange carrier (IC), depending on the time of day, day or week, or distance between the two parties.

2.2.12 Do-Not-Call Enforcement

In the United States, federal and state laws have already mandated do-not-call lists in over half the states, and all states are expected to follow suit. These laws restrict organizations (typically telemarketers) from cold-calling individuals. To comply with these laws, SS7 can be used to query state and federal do-not-call lists (which are stored on a database) each time a telemarketer makes an outbound call. If the number is on a do-not-call list, the call is automatically blocked and an appropriate announcement is played to the marketer. [14]

2.3 The Key to Convergence

Telecommunications network operators can realize increased investment returns by marrying existing SS7/C7 and intelligent networking infrastructures with Internet and other data-centric technologies. SS7/C7 is a key protocol for bridging the telecom and datacom worlds.

The following sections describe the exemplar hybrid network services that SS7/C7 enables:

- ✓ Internet Call Waiting
- ✓ Internet Calling Name Services

- ✓ Click-to-Dial Applications
- ✓ Web-Browser-Based Telecommunication Services
- ✓ WLAN "Hotspot" Billing
- ✓ Location-Based Games

2.3.1 Internet Call Waiting and Internet Calling Name Services

Internet call waiting is a software solution that alerts online Internet users with a call-waiting message on their computer screens when a telephone call enters the same phone line they use for their Internet service. The user can then send the call to voice mail, accept the call, or reject it.

Some providers linking it to CNAM, as mentioned in Calling Name (CNAM), have enhanced the Internet call-waiting service. This service is known as Internet calling name service, and it provides the calling party's name and number.

2.3.2 Click-to-Dial Applications

Click-to-dial applications are another SS7-IP growth area. An example of a click-to-dial application is the ability to click a person's telephone number in an email signature to place a call. These types of services are particularly beneficial to subscribers because they do not require them to change their equipment or access technologies; a POTS and a traditional handset are the only requirements.

2.3.3 Web-Browser-Based of Telecommunication Services

Over the coming decade, we are likely to witness an increase in web based telecommunications services. An example is customer self-provisioning via the Internet, a practice that has been in the marketplace for some time and is likely to increase in both complexity and usage. A customer can already assign himself a premium or toll-free "number for life" via the Internet. The customer can subsequently use a Web interface to change the destination number it points to at will, so that during the day it points to the customer's office phone, and in the evening it points to the customer's cell phone, and so forth.

Another example is the "call me" service, which allows a customer to navigate a Web page to arrange a callback from a department, rather than navigating interactive voice response (IVR) systems through the use of voice prompts and a touch-tone phone.

The marriage is not only between SS7/C7, the Internet, and fixed-line networks—it also extends to cellular networks. Plans are underway to put the location-based information and signaling found in cellular networks into hybrid use. For example, Web-based messenger services could access cellular network home location registers (HLRs) to enable a user to locate a friend or relative in terms of real-time geographic location.

2.3.4 WLAN "Hotspot" Billing

SS7/C7 has recently begun playing a role in the marriage of wireless (WLANs) and cellular networks. A subscriber can use a cellular subscriber identity module (SIM) card for authentication and billing purposes from a WLAN hotspot. For example, if a subscriber is at a café with WLAN facilities (typically Wi-Fi), the subscriber can request permission to use the service via a laptop screen. This request triggers a short cellular call to authenticate the subscriber (using SS7/C7 signaling). The usage is then conveniently billed to the subscriber's cellular phone bill.

2.4 Summary

This chapter has shown that, SS7/C7 plays a very important role in the lives of virtually every individual. At the competitive telecommunication sectors various revenue earning services and QoS perceived by the subscribers is depended vastly on SS7. Furthermore, SS7/C7 is a common thread that ties fixed-line, cellular, and IP networks together and for the telecommunications and data communications industries it is the key enabler.

Chapter 3

SS7 Network Architecture

3.1 Introduction

Before the evolution of SS7 the following systems are used in telecommunication.

CCITT R1 (regional 1) was deployed only on a national level. R1 is a Channel Associated Signaling (CAS) system that was employed in the U.S. and Japan. It uses multifrequency (MF) tones for signaling.

CCITT R2 (regional 2) was deployed only on a national level. R2 is a CAS system that was employed in Europe and most other countries. It used Multifrequency Compelled (MFC) for signaling; it compelled the receiver to acknowledge a pair of tones before sending the next pair.

Signaling systems that have been deployed for both national and international (between international switches) signaling have progressed from CCITT #5 (C5) to CCITT #6 (C6) and finally to CCITT #7 (C7).[2]

    - C5 (CCITT Signaling System No. 5) is a CAS system standardized in 1964 that has found widespread use in international signaling. It is still in use today on a number of international interfaces. National implementations are now scarce, except in less-developed regions of the world, such as Africa, which makes extensive use of the protocol. C5 can be used in both analog and digital environments. In an analog setting, it uses tones for signaling. In a digital setting, a digital representation of the tone is sent instead (a pulse code modulation [PCM] sample).

    - C6 (CCITT Signaling System No. 6), also called SS6, was the first system to employ Common Channel Signaling (CCS). It was standardized in 1972. C6 was a pre-OSI model and as such had a monolithic structure as opposed to a layered one. C6 was a precursor to C7 and included the use of data links to carry signaling in the form of packets. It had error correction/detection mechanisms. It employed a common signaling channel to control a large number of speech circuits, and it had self-governing network management procedures. C6 had a number of advantages over C5, including improvements in post-dial delay and the ability to reject calls with a cause code. The use of locally mapped cause codes allowed international callers to hear announcements in their own language. Although C6 was designed for the international network, it was not as widely deployed as C5. However, it was nationalized for the U.S. network and was deployed quite extensively under the name Common Channel Interoffice Signaling System 6 (CCIS6) in the AT&T network. C6 was introduced into the Bell system in the U.S. in 1976, and soon after, Canada. All deployments have now been replaced by SS7.

3.2 History of SS7

The first specification (called a recommendation by the CCITT/ITU-T) of CCITT Signaling System No. 7 was published in 1980 in the form of the CCITT yellow book recommendations. After the yellow book recommendations, CCITT recommendations were approved at the end of a four-year study period.

Table 3.1- provides an evolutionary time line of CCITT/ITU-T SS7.

Table 3-1: CCITT/ITU-T SS7 Timeline

Year	Publication	Protocols Revised or Added
1980	CCITT Yellow Book	MTP2, MTP3, and TUP, first publication
1984	CCITT Red Book	MTP2, MTP3, and TUP revised. SCCP and ISUP added.
1988	CCITT Blue Book	MTP2, MTP3, TUP, and ISUP revised. ISUP supplementary services and TCAP added.
1992	ITU-T Q.767	International ISUP, first publication.
1993	ITU-T "White Book 93"	ISUP revised.
1996	ITU-T "White Book 96"	MTP3 revised.
1997	ITU-T "White Book 97"	ISUP revised.
1999	ITU-T "White Book 99"	ISUP revised.

Under the CCITT publishing mechanism, the color referred to a published set of recommendations—that is, all protocols were published at the same time. The printed matter had the appropriate colored cover, and the published title contained the color name. When the ITU-T took over from the CCITT, it produced single booklets for each protocol instead of producing en bloc publications as had been the case under the supervision of the CCITT. Under the new mechanism, the color scheme was dropped. As a result, the ITU-T publications came to be known as "White Book" editions, because no color was specified, and the resulting publications had white covers. Because these publications do not refer to a color, you have to qualify the term "White Book" with the year of publication.[2]

As Table 3-1 shows, when SS7 was first published, the protocol stack consisted of only the Message Transfer Part 2 (MTP2), Message Transfer Part 3 (MTP3), and Telephony User Part (TUP) protocols. On first publication, these were still somewhat immature. It was not until the later Red and Blue book editions that the protocol was considered mature. Since then, the SS7 protocols have been enhanced, and new protocols have been added as required.

Figure 3-1 shows how many pages the ITU-T SS7 specifications contained in each year. In 1980, there were a total of 320 pages, in 1984 a total of 641 pages, in 1988 a total of 1900 pages, and in 1999 approximately 9000 pages.

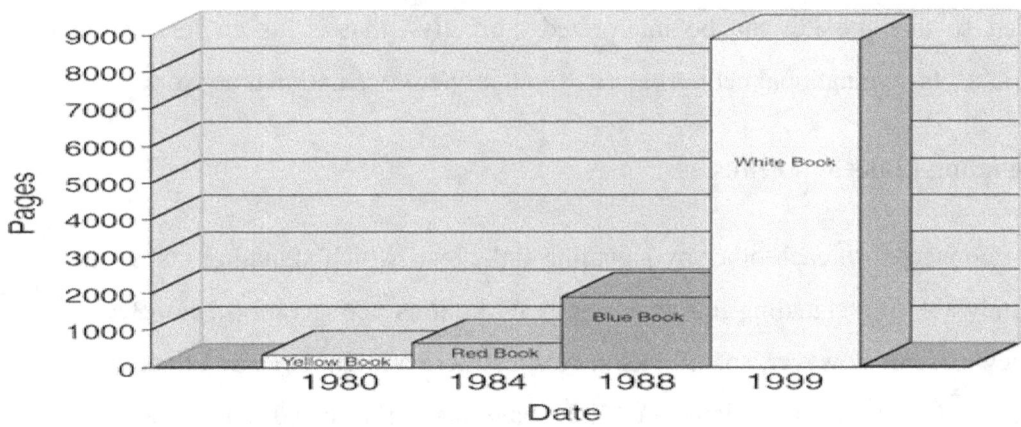

Figure 3-1: How Many Pages the ITU C7 Specifications Covered Based on Year (Source: ITU [Modified])

3.3 SS7 Network Architecture

SS7 can employ different types of signaling network structures. The choice between these different structures can be influenced by factors such as administrative aspects and the structure of the telecommunication network to be served by the signaling system. [2]

The worldwide signaling network has two functionally independent levels:

- International
- National

This structure makes possible a clear division of responsibility for signaling network management. It also lets numbering plans of SS7 nodes belonging to the international network and the different national networks be independent of one another.

SS7 network nodes are called signaling points (SPs). Each SP is addressed by an integer called a point code (PC). The international network uses a 14-bit PC. The national networks also use a 14-bit PC—except North America and China, which use an incompatible 24-bit PC, and Japan, which uses a 16-bit PC. The national PC is unique only within a particular operator's national network. International PCs are unique only within the international network. Other operator networks (if they exist) within a country also could have the same PC and also might share the same PC as that used on the international network. Therefore, additional routing information is provided so that the PC can be interpreted correctly—that is, as an international network, as its own national network, or as another operator's national network

3.4 Signaling Links and Linksets

SPs are connected to each other by signaling links over which signaling takes place. The bandwidth of a signaling link is normally 64 kilobits per second (kbps). Because of legacy reasons, however, some links in North America might have an effective rate of 56 kbps. In recent years, high-speed links have been introduced that use an entire 1.544 Mbps T1 carrier for signaling. Links are typically engineered to carry only 25 to 40 percent of their capacity so that in case of a failure, one link can carry the load of two.

To provide more bandwidth and/or for redundancy, up to 16 links between two SPs can be used. Links between two SPs are logically grouped for administrative and load-sharing reasons. A logical group of links between two SP is called a linkset. Figure 3-2 shows four links in a linkset.

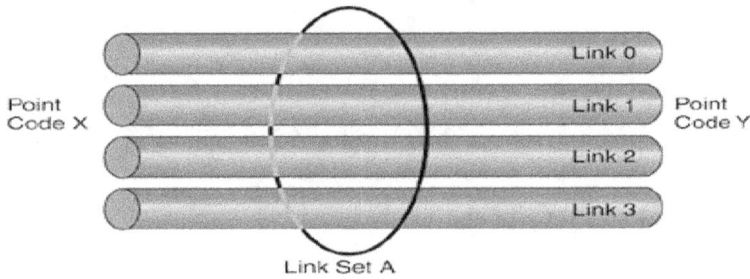

Figure 3-2: Four Links in a Linkset between SPs

A number of linksets that may be used to reach a particular destination can be grouped logically to form a combined linkset. For each combined linkset that an individual linkset is a member of, it may be assigned different priority levels relative to other linksets in each combined linkset.

A group of links within a linkset that have the same characteristics (data rate, terrestrial/satellite, and so on) are called a link group. Normally the links in a linkset have the same characteristics, so the term link group can be synonymous with linkset.

3.5 Routes and Routesets

SS7 routes are statically provisioned at each SP. There are no mechanisms for route discovery. A route is defined as a pre provisioned path between source and destination for a particular relation. Figure 3-3 shows a route from SP A to SP C.

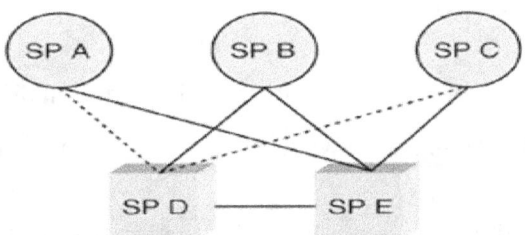

Figure 3-3: Route from SP A to SP C

All the pre provisioned routes to a particular SP destination are called the routeset. Figure 3-4 shows a routeset for SSP C consisting of two routes.

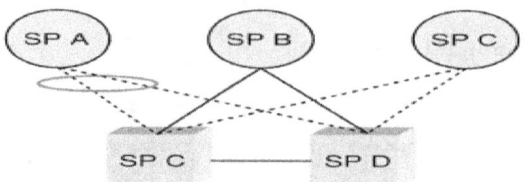

Figure 3-4: Routeset from SP A to SP C

The following section discusses the SP types.

3.6 Node Types

There are three different types of SP (that is, SS7 node):
- Signal Transfer Point
- Service Switching Point
- Service Control Point

Figure 3-5 graphically represents these nodes.

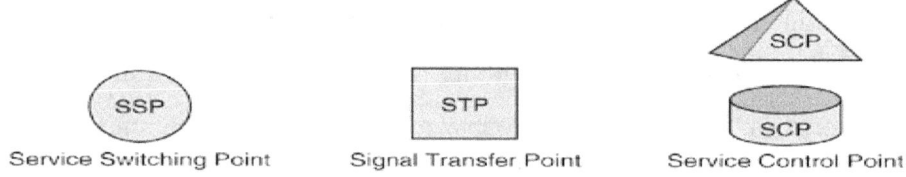

Figure 3-5: SS7 Node Types

The SPs differ in the functions that they perform, as described in the following sections.

3.7 Signal Transfer Point (STP)

A Signal Transfer Point (STP) is responsible for the transfer of SS7 messages between other SS7 nodes, acting somewhat like a router in an IP network.

An STP is neither the ultimate source nor the destination for most signaling messages. Generally, messages are received on one signaling link and are transferred out another. The only messages that are not simply transferred are related to network management and global title translation. STPs route each incoming message to an outgoing signaling link based on routing information contained in the SS7 message. Specifically, this is the information found in the MTP3 routing label. [2]

Additionally, standalone STPs often can screen SS7 messages, acting as a firewall. An STP can exist in one of two forms:

- Standalone STP
- Integrated STP (SP with STP)

Standalone STPs are normally deployed in "mated" pairs for the purposes of redundancy. Under normal operation, the mated pair shares the load. If one of the STPs fails or isolation occurs because of signaling link failure, the other STP takes the full load until the problem with its mate has been rectified.

Integrated STPs combine the functionality of an SSP and an STP. They are both the source and destination for MTP user traffic. They also can transfer incoming messages to other nodes.

There are three levels of STPs.

- National Signal Transfer Point
- International Signal Transfer Point
- Gateway Signal Transfer Point

3.7.1 National STP

A National STP exists within the national network (will vary with the country). It can transfer messages that use the same national standard of protocol.
Messages can be passed to an International STP, but cannot be converted by the National STP. Protocol converters often interconnect a National and an International STP by converting from ANSI to ITU-TS.

3.7.2 International STP

An International STP functions within an international network. It provides for SS7 interconnection of all countries, using the ITU-TS standard protocol.
All nodes connecting to an International STP must use the ITU-TS protocol standard.

3.7.3 Gateway STP

A Gateway STP converts signaling data from one protocol to another. Gateway STPs are often used as an access point to the international network. National protocols are converted to the ITU-TS protocol standard. Depending on its location, the Gateway STP must be able to use both the International and National protocol standards.

A Gateway STP also serves as an interface into another network's databases, such as from an interexchange carrier (IXC) to an end office. The Gateway STP can also be configured to screen for authorized users of the network.

Gateway STPs also provide measurements of traffic and usage via the following means:

- **Traffic**—Measures the peg counts of the type of messages entering or leaving the network.
- **Network events**—Track events such as link out-of-service or local processor outage, for maintenance purposes.
- **Usage**—Provides peg counts of the record number of messages by message type. Usage counts are sent to the Regional Accounting Office (RAO) for processing in Bell Networks. RAOs invoice customers such as IXCs and independent telcos, charging for access into the SS7 network, to help offset the cost of deploying the network.

3.8 Service Switching Point (SSP)

A Service Switching Point (SSP) is a voice switch that incorporates SS7 functionality. It processes voice-band traffic (voice, fax, modem, and so forth) and performs SS7 signaling. All switches with SS7 functionality are considered SSPs regardless of whether they are local switches (known in North America as an end office) or tandem switches.[2]

An SSP can originate and terminate messages, but it cannot transfer them. If a message is received with a point code that does not match the point code of the receiving SSP, the message is discarded.

3.9 Service Control Point (SCP)

A Service Control Point (SCP) acts as an interface between telecommunications databases and the SS7 network. Telephone companies and other telecommunication service providers employ a number of databases that can be queried for service data for the provision of services. Typically the request (commonly called a query) originates at an SSP. A popular example is freephone calling (known as toll-free in North America). The SCP provides the routing number (translates the toll-free number to a routable number) to the SSP to allow the call to be completed.

SCPs form the means to provide the core functionality of cellular networks, which is subscriber mobility. Certain cellular databases (called registers) are used to keep track of the subscriber's location so that incoming calls may be delivered. Other telecommunication databases include those used for calling card validation (access card, credit card), calling name display (CNAM), and LNP.

SCPs used for large revenue-generating services are usually deployed in pairs and are geographically separated for redundancy. Unless there is a failure, the load is typically shared between two mated SCPs. If failure occurs in one of the SCPs, the other one should be able to take the load of both until normal operation resumes.
Queries/responses are normally routed through the mated pair of STPs that services that particular SCP, particularly in North America.

3.10 Link Types

Signaling links can be referenced differently depending on where they are in the network. Although different references can be used, you should understand that the link's physical characteristics remain the same. The references to link types A through E are applicable only where standalone STPs are present, so the references are more applicable to the North American market.

Six different link references exist:
- Access links (A links)
- Crossover links (C links)
- Bridge links (B links)
- Diagonal links (D links)
- Extended links (E links)
- Fully associated links (F links)

The following sections cover each link reference in more detail.

3.10.1 Access Links (A Links)

Access links (A links), shown in Figure 3-6. provide access to the network. They connect "outer" SPs (SSPs or SCPs) to the STP backbone. A links connect SSPs and SCPs to their serving STP or STP mated pair.

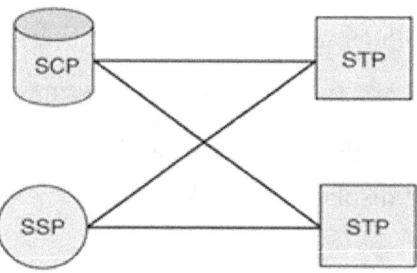

Figure 3-6: A Links

3.10.2 Cross Links (C Links)

Cross links (C links), shown in Figure 3-7, are used to connect two STPs to form a mated pair—that is, a pair linked such that if one fails, the other takes the load of both.

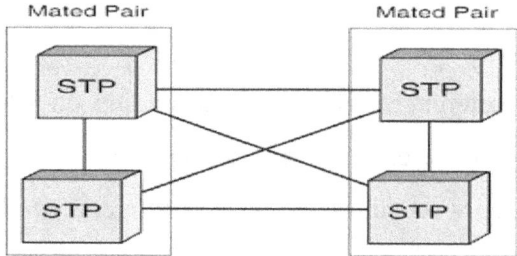

Figure 3-7: C Links

C links are used to carry MTP user traffic only when no other route is available to reach an intended destination. Under normal conditions, they are used only to carry network management messages.

3.10.3 Bridge Links (B Links)

Bridge links (B links) are used to connect mated pairs of STPs to each other across different regions within a network at the same hierarchical level. These links help form the backbone of the SS7 network. B links are normally deployed in link quad configuration between mated pairs for redundancy. Figure 3-8 shows two sets of mated pairs of B links.

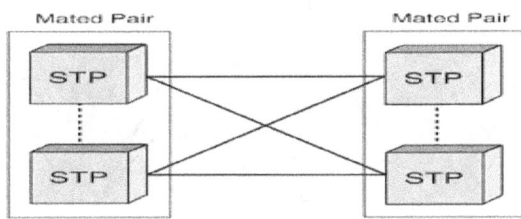

Figure 3-8: B Links

3.10.4 Diagonal Links (D Links)

Diagonal links (D links), shown in Figure 3-9, are the same as B links in that they connect mated STP pairs.

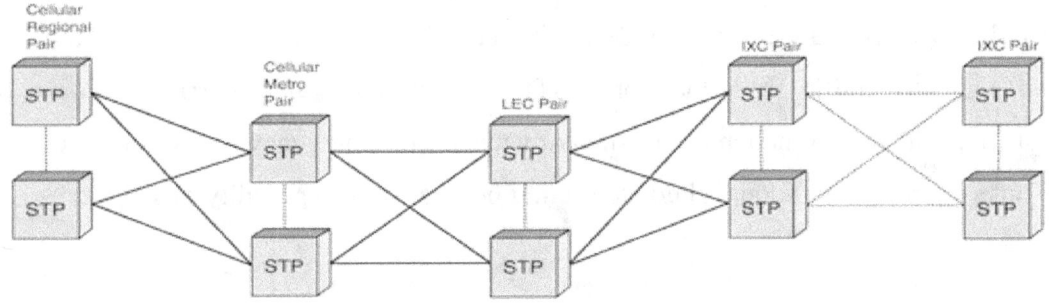

Figure 3-9: D Links

The difference is that they connect mated STP pairs that belong to different hierarchical levels or to different networks altogether. For example, they may connect an interexchange carrier (IXC) STP pair to a local exchange carrier (LEC) STP pair or a cellular regional STP pair to a cellular metro STP pair.

As mentioned, B and D links differ in that D links refer specifically to links that are used either between different networks and/or hierarchical levels, as shown in Figure 3-10.

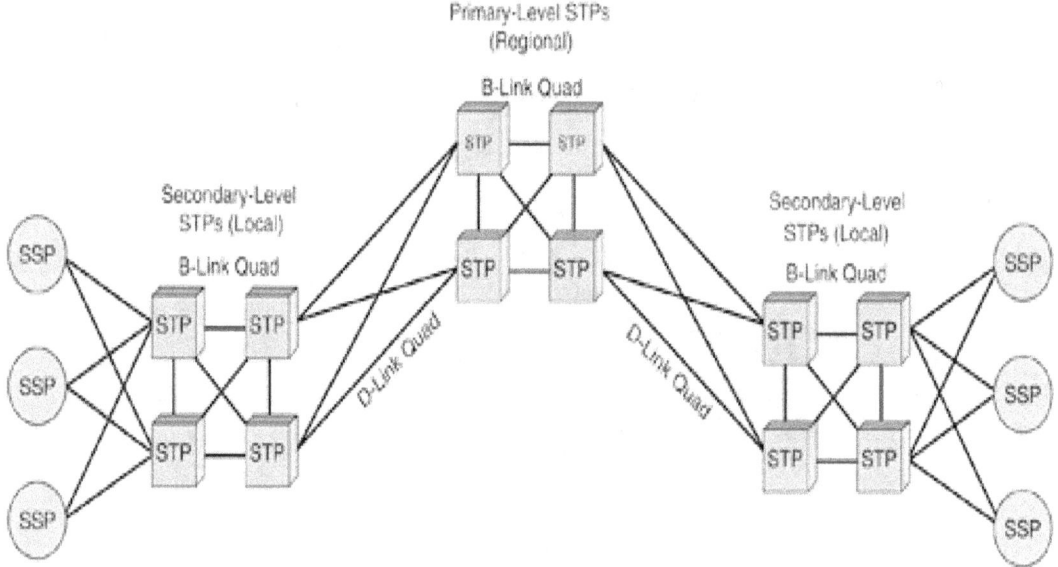

Figure 3-10: Existence of an STP Backbone and STP Hierarchy

3.10.5 Extended Links (E Links)

Extended links (E links), shown in Figure 3-11, connect SSPs and SCPs to an STP pair, as with A links, except that the pair they connect to is not the normal home pair. Instead, E links connect to a non home STP pair. They are also called alternate access (AA) links. E links are used to provide additional reliability or, in some cases, to offload signaling traffic from the home STP pair in high-traffic corridors. For example, an SSP serving national government agencies or emergency services might use E links to provide additional alternate routing because of the criticality of service.

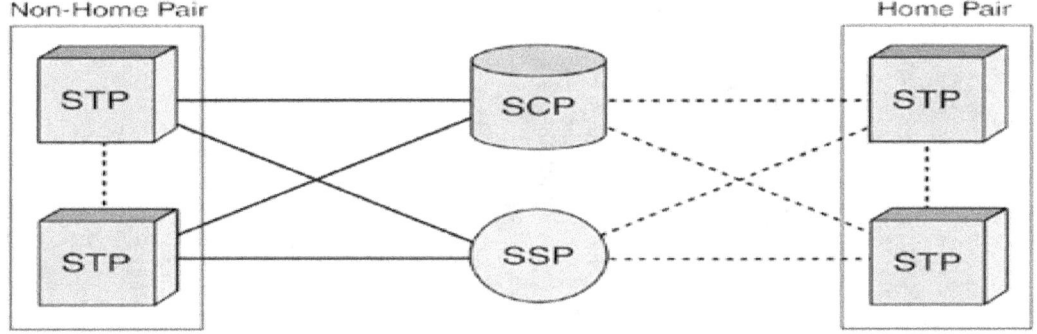

Figure 3-11: E Links

3.10.6 Fully-Associated Links (F Links)

Fully-associated links (F links), shown in Figure 3-12, are used to connect network SSPs and/or SCPs directly to each other without using STPs. The most common application of this type of link is in metropolitan areas. F links can establish direct connectivity between all switches in the area for trunk signaling and Custom Local Area Signaling Service (CLASS), or to their corresponding SCPs.

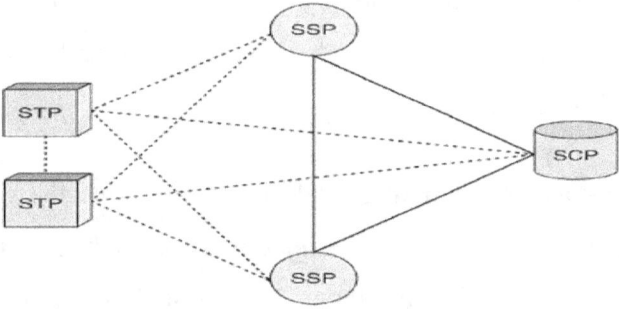

Figure 3-12: F Links

Figure 3-13 shows an SS7 network segment. In reality, there would be several factors more SSPs than STPs.

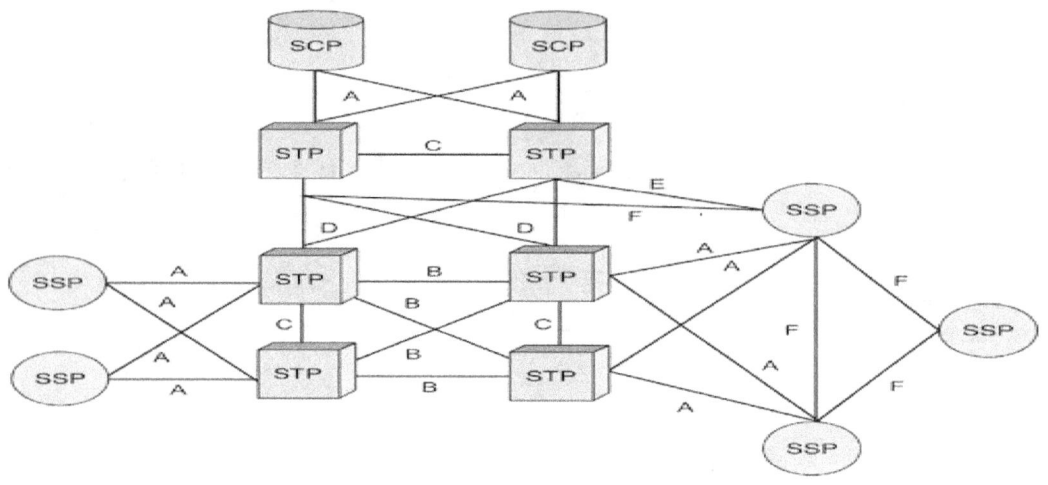

Figure 3-13: SS7 Network Segment

3.11 Signaling Modes

The signaling relationship that exists between two communicating SS7 nodes is called the signaling mode. The two modes of signaling are associated signaling and quasi-associated signaling. When the destination of an SS7 message is directly connected by a linkset, the associated signaling mode is being used. In other words, the source and destination nodes are directly connected by a single linkset. When the message must pass over two or more linksets and through an intermediate node, the quasi-associated mode of signaling is being used.[2]

It's easier to understand the signaling mode if you examine the relationship of the point codes between the source and destination node. When using the associated mode of signaling, the Destination Point Code (DPC) of a message being sent matches the PC of the node at the far end of the linkset, usually referred to as the far-end PC or adjacent PC. When quasi-associated signaling is used, the DPC does not match the PC at the far end of the connected linkset. Quasi-associated signaling requires the use of an STP as the intermediate node because an SSP cannot transfer messages.

In Figure 4-14, the signaling relationships between each of the nodes are as follows:

- SSP A to SSP B uses quasi-associated signaling.
- SSP B to SSP C uses associated signaling.
- STP 1 and STP 2 use associated signaling to SSP A, SSP B, and each other.

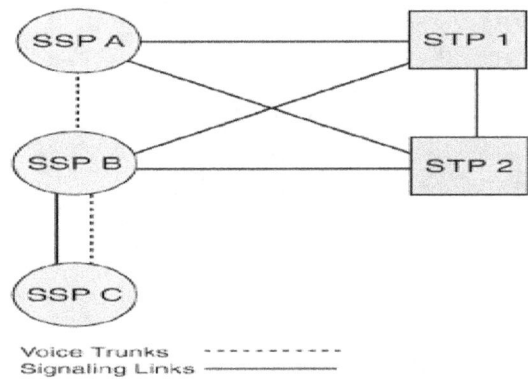

Figure 3-14: SS7 Signaling Modes

As you can see from Figure 3-14, associated signaling is used between nodes that are directly connected by a single linkset, and quasi-associated signaling is used when an intermediate node is used. Notice that SSP C is only connected to SSP B using an F link. It is not connected to any other SS7 nodes in the figure.

When discussing the signaling mode in relation to the voice trunks shown between the SSPs, the signaling and voice trunks follow the same path when associated signaling is used. They take separate paths when quasi-associated signaling is used. You can see from Figure 3-15 that the signaling between SSP B and SSP C follows the same path (associated mode) as the voice trunks, while the signaling between SSP A and SSP B does not follow the same path as the voice trunks.

3.12 Signaling Network Structure

Standalone STPs are prevalent in North America because they are used in this region to form the backbone of the SS7 network. Attached to this backbone are the SSPs and SCPs. Each SSP and SCP is assigned a "home pair" of STPs that it is directly connected to. The network of STPs can be considered an overlay onto the telecommunications network—a packet-switched data communications network that acts as the nervous system of the telecommunications network.

Figure 3-15 shows a typical example of how SSPs are interconnected with the STP network in North America.

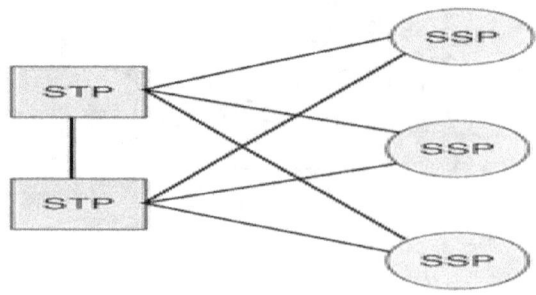

Figure 3-15: Typical Example of North American SSP Interconnections.

STPs are not as common outside North America. Standalone STPs typically are used only between network operators and/or for applications involving the transfer of non circuit-related signaling. In these regions, most SSPs have direct signaling link connections to other SSPs to which they have direct trunk connections.

Figure 3-16 shows an example of this type of network with most SSPs directly connected by signaling links.

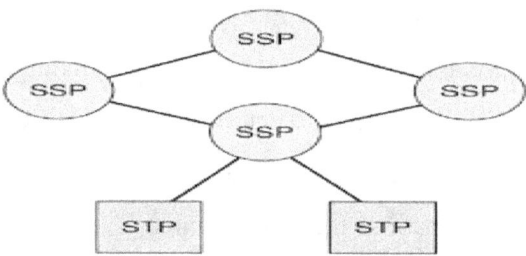

Figure 3-16: Typical Example of SSP Interconnections in Most Areas Outside North America

SSPs often have indirect physical connections to STPs, made through other SSPs in the network. These are usually implemented as nailed-up connections, such as through a Digital Access Cross-Connect System or other means of establishing a semi permanent connection. Logically, these SSPs are directly connected to the STP. The signaling link occupies a digital time slot on the same physical medium as the circuit-switched traffic. The SSPs that provide physical interconnection between other SSPs and an STP do not "transfer" messages as an STP function. They only provide physical connectivity of the signaling links between T1/E1 carriers to reach the STP.[2]

Figure 3-17 shows an example of a network with no STP connection, direct connections, and non direct connections. SSP 1 is directly connected to an STP pair. SSP 4 uses direct signaling links to SSP 2 and SSP 3, where it also has direct trunks. It has no STP connection at all. SSP 2 and SSP 3 are connected to the STP pair via nailed-up connections at SSP 1.

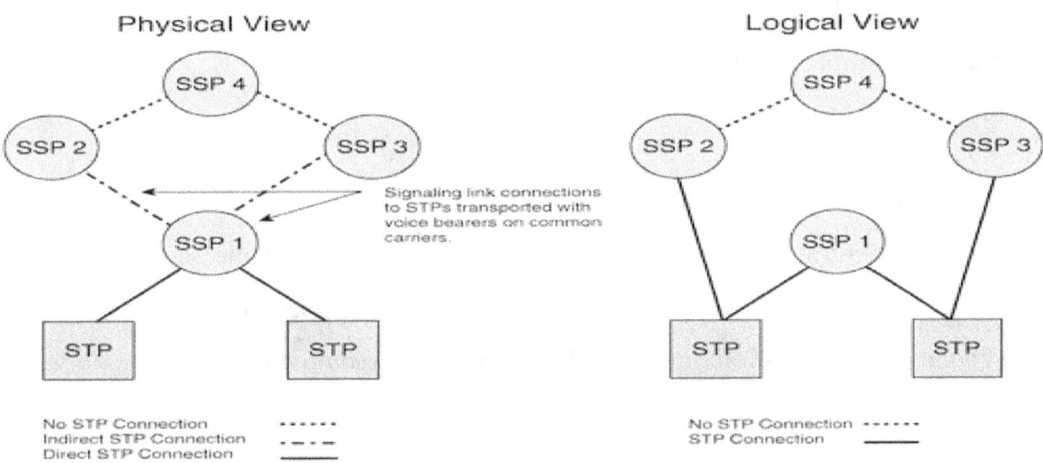

Figure 3-17: Example of Direct and Indirect SSP Interconnections to STPs

Normally within networks that do not use STPs, circuit-related (call-related) signaling takes the same path through the network as user traffic because there is no physical need to take a different route. This mode of operation is called associated signaling and is prevalent outside North America. Referring back to Figure 3-14, both the user traffic and the signaling take the same path between SSP B and SSP C.[2]

Because standalone STPs are used to form the SS7 backbone within North America, and standalone STPs do not support user traffic switching, the SSP's signaling mode is usually quasi-associated, as illustrated between SSP A and SSP B in Figure 3-14.

In certain circumstances, the SSP uses associated signaling within North America. A great deal of signaling traffic might exist between two SSPs, so it might make more sense to place a signaling link directly between them rather than to force all signaling through an STP.

3.13 Summary

SS7 is a data communications network that acts as the nervous system to bring the components of telecommunications networks to life. It acts as a platform for various services described throughout this chapter. SS7 nodes are called signaling points (SPs), of which there are three types:

- Service Switching Point (SSP)
- Service Control Point (SCP)
- Signal Transfer Point (STP)

SSPs provide the SS7 functionality of a switch. STPs may be either standalone or integrated STPs (SSP and STP) and are used to transfer signaling messages. SCPs interface the SS7 network to query telecommunication databases, allowing service logic and additional routing information to be obtained to execute services.

Chapter 4

Protocols of SS7

4.1 SS7 Protocol Overview

The number of possible protocol stack combinations is growing. It depends on whether SS7 is used for cellular-specific services or intelligent network services, whether transportation is over IP or is controlling broadband ATM networks instead of time-division multiplexing (TDM) networks, and so forth. This requires coining a new term—traditional SS7—to refer to a stack consisting of the protocols widely deployed from the 1980s to the present:

- Message Transfer Parts (MTP 1, 2, and 3)
- Signaling Connection Control Part (SCCP)
- Transaction Capabilities Application Part (TCAP)
- Telephony User Part (TUP)
- ISDN User Part (ISUP)

Figure 4-1 shows a common introductory SS7 stack.

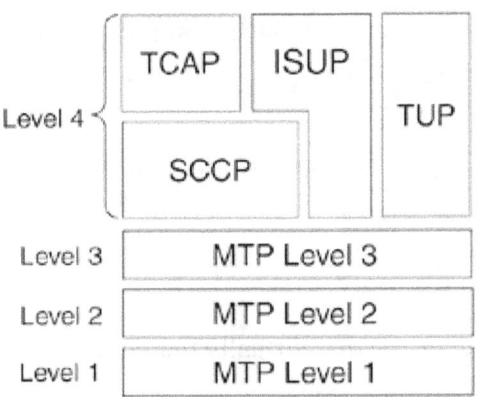

Figure 4-1: Introductory SS7 Protocol Stack

Such a stack uses TDM for transport. This book focuses on traditional SS7 because that is what is implemented. Newer implementations are beginning to appear that use different transport means such as IP and that have associated new protocols to deal with the revised transport.

The SS7 physical layer is called MTP level 1 (MTP1), the data link layer is called MTP level 2 (MTP2), and the network layer is called MTP level 3 (MTP3). Collectively they are called the Message Transfer Part (MTP). The MTP protocol is SS7's native means of packet transport. In recent years there has been an interest in the facility to transport SS7 signaling over IP instead of using SS7's native MTP. This effort has largely been carried out by the Internet Engineering Task Force (IETF) SigTran (Signaling Transport) working group.[2]

TUP and ISUP both perform the signaling required to set up and tear down telephone calls. As such, both are circuit-related signaling protocols. TUP was the first call control protocol specified. It could support only plain old telephone service (POTS) calls. Most countries are replacing TUP with ISUP. Both North America and Japan bypassed TUP and went straight from earlier signaling systems to ISUP. ISUP supports both POTS and ISDN calls. It also has more flexibility and features than TUP.

4.2 SS7 Protocol Stack

This chapter describes the components of the SS7 protocol stack. A stack is a set of data storage locations that are accessed in a fixed sequence. The SS7 stack is compared against the Open Systems Interconnection (OSI) model for communication between different systems made by different vendors.[15]

Figure 4-2 shows the components of the SS7 protocol stack.

4.2.1 SS7 Level 1: Physical Connection

This is the physical level of connectivity, virtually the same as Layer 1 of the OSI model. SS7 specifies what interfaces will be used, both Bellcore (Telecordia) and ANSI call for either the DS0A or the V.35 interface.[15]
Because central offices are already using DS1 and DS3 facilities to link one another, the DS0A interface is readily available in all central offices, and is preferred in the SS7 network. As the demands on the SS7 network increase (local number portability), and as the industry migrates toward ATM networks, the DS1 interface will become the link interface.

4.2.2 SS7 Level 2: Data Link

The data link level provides the network with sequenced delivery of all SS7 message packets. Like the OSI data link layer, it is only concerned with the transmission of data from one node to the next, not to its final destination in the network. Sequential numbering is used to determine if any messages have been lost during transmission. Each link uses its own message numbering series independent of other links.[15]

SS7 uses CRC-16 error checking of data and requests retransmission of lost or corrupted messages. Length indicators allow Level 2 to determine what type of signal unit it is receiving, and how to process it.

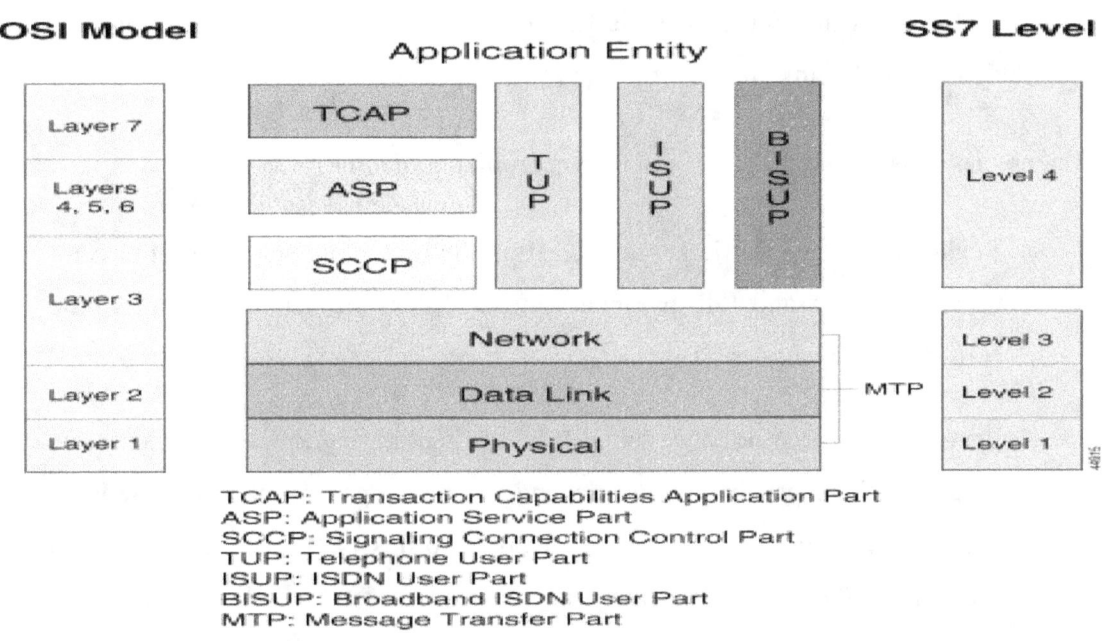

TCAP: Transaction Capabilities Application Part
ASP: Application Service Part
SCCP: Signaling Connection Control Part
TUP: Telephone User Part
ISUP: ISDN User Part
BISUP: Broadband ISDN User Part
MTP: Message Transfer Part

Figure 4-2: SS7 Protocol Stack

4.2.3 SS7 Level 3: Network Level

The network level depends on the services of Level 2 to provide routing, message discrimination and message distribution functions.

- Message Discrimination determines to whom the message is addressed.
- Message Distribution is passed here if it is a local message.
- Message Routing is passed here if it is not a local message.

Message Discrimination

This function determines whether a message is local or remote using the point code and data contained in a lookup table. Messages to remote destinations are passed to the message routing function for additional processing.

Message Distribution

Message distribution provides link, route and traffic management functions.

Link Management

This function uses the Link Status Signal Unit (LSSU) to notify adjacent nodes of link problems. Level 3 will send LSSUs via Level 2 to the adjacent node, notifying it of the problems with the link and its status.[15]

Diagnostics consists of realigning and re synchronizing the link.

- Realignment—All traffic is removed from the link, counters are reset to zero, timers are reset and Fill-In Signal Units (FISUs) are sent in the meantime (called the proving period).

- Proving Period—Amount of time FISUs are sent during link realignment. The duration of the proving period depends on the type of link used. Bellcore specifies the proving period for a 56 Kbps DS0 link is 2.3 seconds for normal proving and 0.6 seconds for emergency proving.

Another form of link management uses changeover and changeback messages sent using Message Signal Units (MSUs). MSUs advise the adjacent node to send traffic over another link within the same linkset. The alternate link must be within the same linkset.

The bad link is being realigned by Level 3 while traffic is rerouted over alternate links. Changeback message is sent to advise the adjacent node that it can use the newly restored link again. Changeback messages are typically followed by a changeback acknowledgement message.

Route Management

This function provides a means for rerouting traffic around failed or congested nodes. Route management is a function of Level 3 and works together with link management.

Route management informs other nodes of the status of the affected node. It uses Message Signal Units (MSUs) generated by adjacent nodes and is not usually generated by the affected nodes. (Link management only informs adjacent nodes.)

Traffic Management

This function provides flow control if a node has become congested. It allows the network to control the flow of certain messages based on protocol. Traffic management deals with a specific user part within an affected node.[15]

For example, if ISUP is not available at a particular node, a traffic management message can be sent to adjacent nodes informing them that ISUP is not available, without affecting TCAP messages on the same node.

Message Routing

Message discrimination in Level 3 will pass messages to message routing if it determines the message is not local. Message routing reads the called and calling party addresses to determine the physical address in the form of a point code.[15]

Every SS7 node must have its own unique point code. Message routing determines the point code from an address contained in the routing table.

4.3 MTP

MTP levels 1 through 3 are collectively referred to as the MTP. The MTP comprises the functions to transport information from one SP to another.

The MTP transfers the signaling message, in the correct sequence, without loss or duplication, between the SPs that make up the SS7 network. The MTP provides reliable transfer and delivery of signaling messages. The MTP was originally designed to transfer circuit-related signaling because no noncircuit-related protocol was defined at the time.[2]

The recommendations refer to MTP1, MTP2, and MTP3 as the physical layer, data link layer, and network layer, respectively. The following sections discuss MTP2 and MTP3. (MTP1 isn't discussed because it refers to the physical network.)

MTP2

Signaling links are provided by the combination of MTP1 and MTP2. MTP2 ensures reliable transfer of signaling messages. It encapsulates signaling messages into variable-length SS7 packets. SS7 packets are called signal units (SUs). MTP2 provides delineation of SUs, alignment of SUs, signaling link error monitoring, error correction by retransmission, and flow control. The MTP2 protocol is specific to narrowband links (56 or 64 kbps).

MTP3

MTP3 performs two functions:

- **Signaling Message Handling (SMH)**— Delivers incoming messages to their intended User Part and routes outgoing messages toward their destination. MTP3 uses the PC to identify the correct node for message delivery. Each message has both an Origination Point Code (OPC) and a DPC. The OPC is inserted into messages at the MTP3 level to identify the SP that originated the message. The DPC is inserted to identify the address of the destination SP. Routing tables within an SS7 node are used to route messages.[2]
- **Signaling Network Management (SNM)**— Monitors linksets and routesets, providing status to network nodes so that traffic can be rerouted when necessary. SNM also provides procedures to take corrective action when failures occur, providing a self-healing mechanism for the SS7 network.

Figure 4-3 shows the relationship between levels 1, 2, and 3

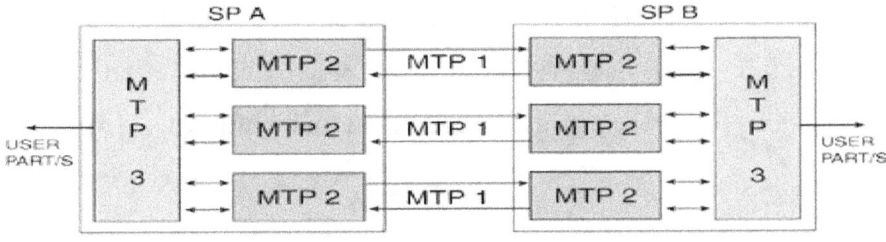

Figure 4-3: A Single MTP3 Controls Many MTP2s, Each of Which Is Connected to a Single MTP1

4.4 User and Application Parts

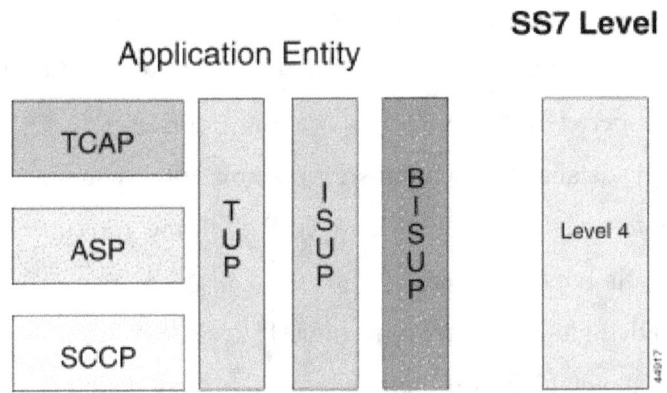

Figure 4-4: SS7 Level 4 Protocols, User and Application Parts[15]

4.4.1 TUP and ISUP

TUP and ISUP sit on top of MTP to provide circuit-related signaling to set up, maintain, and tear down calls. TUP has been replaced in most countries because it supports only POTS calls. Its successor, ISUP, supports both POTS and ISDN calls as well as a host of other features and added flexibility. Both TUP and ISUP are used to perform interswitch call signaling. ISUP also has inherent support for supplementary services, such as automatic callback, calling line identification, and so on.[2]

4.4.2 BISUP

Broadband ISDN User Part (BISUP) is an ATM protocol intended to support services such as high-definition television (HDTV), multilingual TV, voice and image storage and retrieval, video conferencing, high-speed LANs and multimedia.[15]

4.5 SCCP

The combination of the MTP and the SCCP is called the Network Service Part (NSP) in the specifications (but outside the specifications, this term is seldom used).

The addition of the SCCP provides a more flexible means of routing and provides mechanisms to transfer data over the SS7 network. Such additional features are used to support non circuit-related signaling, which is mostly used to interact with databases (SCPs). It is also used to connect the radio-related components in cellular networks and for inter-SSP communication supporting CLASS services. SCCP also provides application management functions. Applications are mostly SCP database driven and are called subsystems. For example, in cellular networks, SCCP transfers queries and responses between the Visitor Location Register (VLR) and Home Location Register (HLR) databases. Such transfers take place for a number of reasons. The primary reason is to update the subscriber's HLR with the current VLR serving area so that incoming calls can be delivered.[2]

Enhanced routing is called global title (GT) routing. It keeps SPs from having overly large routing tables that would be difficult to provision and maintain. A GT is a directory number that serves as an alias for a physical network address. A physical address consists of a point code and an application reference called a subsystem number (SSN). GT routing allows SPs to use alias addressing to save them from having to maintain overly large physical address tables. Centralized STPs are then used to convert the GT address into a physical address; this process is called Global Title Translation (GTT). This provides the mapping of traditional telephony addresses (phone numbers) to SS7 addresses (PC and/or SSN) for enhanced services. GTT is typically performed at STPs.

4.6 TCAP

TCAP allows applications (called subsystems) to communicate with each other (over the SS7 network) using agreed-upon data elements. These data elements are called components. Components can be viewed as instructions sent between applications. For example, when a subscriber changes VLR location in a global system for mobile communication (GSM) cellular network, his or her HLR is updated with the new VLR location by means of an UpdateLocation component. TCAP also provides transaction management, allowing multiple messages to be associated with a particular communications exchange, known as a transaction.[2]

There are a number of subsystems; the most common are
- Toll-free (E800)
- Advanced Intelligent Network (AIN)
- Intelligent Network Application Protocol (INAP)
- Customizable Applications for Mobile Enhanced Logic (CAMEL)

4.7 MAP

The Mobile Application Part (MAP) is an SS7 protocol which provides an application layer for the various nodes in GSM and UMTS mobile core networks and GPRS core networks to communicate with each other in order to provide services to mobile phone users. The Mobile Application Part is the application-layer protocol used to connect the distributed switching elements, called mobile switching centers (MSCs) with a master database called the Home Location Register (HLR). The HLR dynamically stores the current location and profile of a mobile network subscriber. The HLR is consulted during the processing of an incoming call. Conversely, the HLR is updated as the subscriber moves about the network and is thus serviced by different switches within the network. [15]

4.8 Summary

Signaling is transferred using the packet-switching facilities afforded by SS7. These packets are called signal units (SUs). The Message Transfer Part (MTP) and the Signaling Connection Control Part (SCCP) provide the transfer protocols. MTP is used to reliably transport messages between nodes, and SCCP is used for noncircuit-related signaling (typically, transactions with SCPs). The ISDN User Part (ISUP) is used to set up and tear down both ordinary (analog subscriber) and ISDN calls. The Transaction Capabilities Application Part (TCAP) allows applications to communicate with each other using agreed-upon data components and manages transactions.

Chapter 5

Message Transfer Part (MTP)

5.1 Introduction to MTP

MTP constitutes the bottom layers (LI, L2 and L3) of SS7, providing physical links for signaling transmission to ensure reliable message transfer. MTP also provides signaling route management and signaling network management functions. Figure 5-1 shows the position of the MTP in SS7. [16]

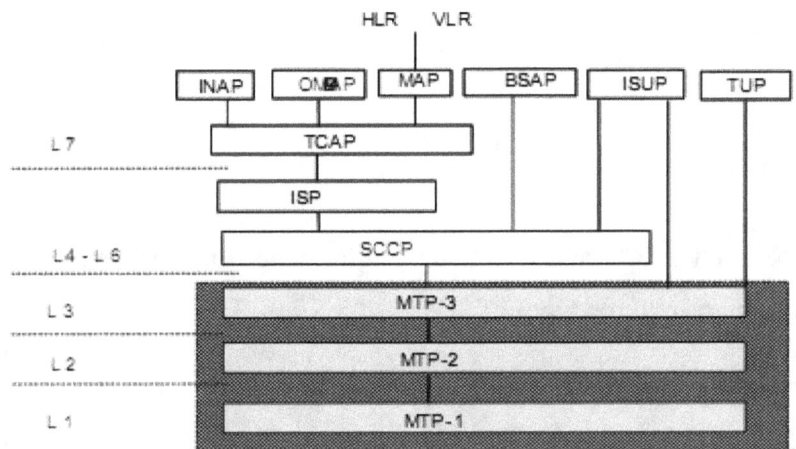

Figure 5-1: Position of the MTP in SS7

5.2 Different types of MTP

The Message Transfer Part (MTP) is part of the Signaling System 7 (SS7) used for communication in Public Switched Telephone Networks. MTP is responsible for reliable, unduplicated and in-sequence transport of SS7 messages between communication partners.

- MTP Level 1 (Physical layer)
- MTP Level 2 (Data link layer)
- MTP Level 3 (Network layer)

5.2.1 MTP 1 Functions.

MTP Level 1 (Physical layer) defines the physical, electrical, and functional characteristics of the digital signaling link. Physical interfaces defined include **E-1** (2048 kb/s; 32 64 kb/s channels), **DS-1** (1544 kb/s; 24 64 kp/s channels), **V.35** (64 kb/s), **DS-0** (64 kb/s), and **DS-0A** (56 kb/s).

That is, the layer that is responsible for the connection of SS7 Signaling Points into the transmission network over which they communicate with each other. Primarily, this involves the conversion of messaging into electrical signal and the maintenance of the physical links through which these pass. In this way, it is analogous to the Layer 1 of ISDN or other, perhaps more familiar, protocols.

5.2.2 MTP 2 Functions.

MTP2 works point to point and "frames" signaling information into packets called signaling units (SUs). There are three types of SUs:

- Fill-in Signal Unit (FISU)
- Link Status Signal Unit (LSSU)
- Message Signal Unit (MSU)

MTP2 uses flags (delimitation) to separate SUs.

FISUs are fillers that are sent when no LSSUs or MSUs are to be sent. LSSUs are sent to convey link status information between two adjacent signaling points (SPs). MSUs carry the real signaling content: messages for call control, network management, and TCAP query/response.

5.2.2.1 Fill-In Signal Units

FISUs are the most basic SU and carry only MTP2 information. They are sent when there are no LSSUs or MSUs to be sent, when the signaling link would otherwise be idle. Sending FISUs ensures 100 percent link occupancy by SUs at all times. A cyclic redundancy check (CRC) checksum is calculated for each FISU, allowing both signaling points at either end of the link to continuously check signaling link quality. [2]

The seven fields that comprise a FISU, shown in Figure 5-2 are also common to LSSUs and MSUs.

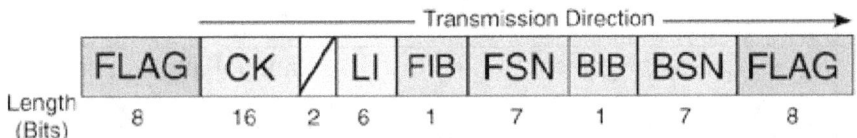

Figure 5-2: FISU Format

5.2.2.2 Link Status Signal Units

LSSUs carry one or two octets of link status information between signaling points at either end of a link. The link status controls link alignment, indicates the link's status, and indicates a signaling point's status to the remote signaling point. The presence of LSSUs at any time other than during link alignment indicates a fault—such as a remote processor outage or an unacceptably high bit error rate affecting the ability to carry traffic.

The timers associated with a particular status indication govern the transmission interval. After the fault is cleared, the transmission of LSSUs ceases, and normal traffic flow can continue. As with FISUs, only MTP2 of adjacent signaling points exchanges LSSUs. LSSUs are identical to FISUs, except that they contain an additional field called the Status field (SF). Figure 5-3 shows the eight fields that comprise an LSSU.

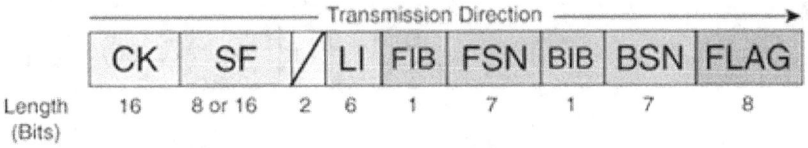

Figure 5-3: LSSU Format

Currently only a single-octet SF is used, even though the specifications allow for a two-octet SF. From the single octet, only the first 3 bits are defined. These bits provide the status indications shown in Table 5-1

Table 5-1: Values in the Status Field

C	B	A	Status Indication	Acronym	Meaning
0	0	0	O: Out of Alignment	SIO	Link not aligned; attempting alignment
0	0	1	N: Normal Alignment	SIN	Link is aligned
0	1	0	E: Emergency Alignment	SIE	Link is aligned
0	1	1	OS: Out of Service	SIOS	Link out of service; alignment failure
1	0	0	PO: Processor Outage	SIPO	MTP2 cannot reach MTP3
1	0	1	B: Busy	SIB	MTP2 congestion

5.2.2.3 Message Signal Units

As shown in Figure 5-4 MSUs contain the common fields of the FISU and two additional fields: the Signaling Information Field (SIF) and the Service Information Octet (SIO). MSUs carry the signaling information (or messages) between both MTP3 and Level 4 users. The messages include all call control, database query, and response messages. In addition, MSUs carry MTP3 network management messages. All messages are placed in the SIF of the MSU. [2]

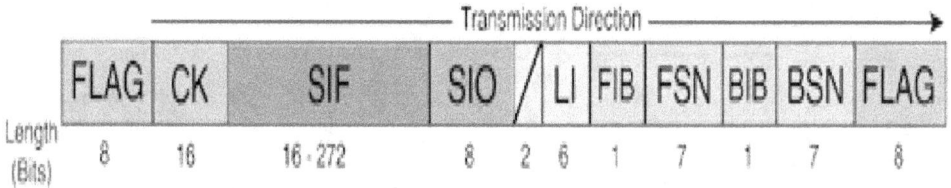

Figure 5-4: MSU Format

5.2.2.4 Field Descriptions

Table 5-2 details the fields that are found inside the signal units. MTP2 exclusively processes all fields except the SIO and the SIF.

Table 5-2: Field Descriptions

Field	Length in Bits	Description
Flag	8	A pattern of 01111110 to indicate the start and end of an SU.
BSN	7	Backward sequence number. Identifies the last correctly received SU.
BIB	1	Backward indicator bit. Toggled to indicate an error with the received SU.
FSN	7	Forward sequence number. Identifies each transmitted SU.
FIB	1	Forward indicator bit. Toggled to indicate the retransmission of an SU that was received in error by the remote SP.
LI	6	Length indicator. Indicates how many octets reside between itself and the CRC field. The LI field also implies the type of signal unit. LI = 0 for FISUs, LI = 1 or 2 for LSSUs, and LI >2 for MSUs.
SF	8 to 16	Status field. Provides status messages in the LSSU only.
CK	16	Check bits. Uses CRC-16 to detect transmission errors.
SIO	8	Service Information Octet. Specifies which MTP3 user has placed a message in the SIF.
SIF	16 to 2176	Signaling Information Field. Contains the "real" signaling content. The SIF is also related to call control, network management, or databases query/response.

5.2.2.5 Signal Unit Delimitation

A flag octet that is coded as 01111110 separates consecutive signal units on a signaling data link. The flag octet indicates the beginning or end of an SU.

NOTE

It is optional whether a single flag is used to mark both the beginning and end of an SU, or whether a common flag is used for both. The latter is the most common implementation.

Because the 01111110 flag pattern can also occur in an SU, the SU is scanned before a flag is attached, and a 0 is inserted after every sequence of five consecutive 1s. This method is called bit stuffing (or 0 bit insertion). It solves the problem of false flags, because it prevents the pattern 01111110 from occurring inside an SU. The receiving MTP2 carries out the reverse process, which is called bit removal (or 0 bit deletion).

After flag detection and removal, each 0 that directly follows a sequence of five consecutive 1s is deleted. Figure 5-5 shows how the sending node adds a 0 following five 1s while the receiving node removes a 0 following five 1s.

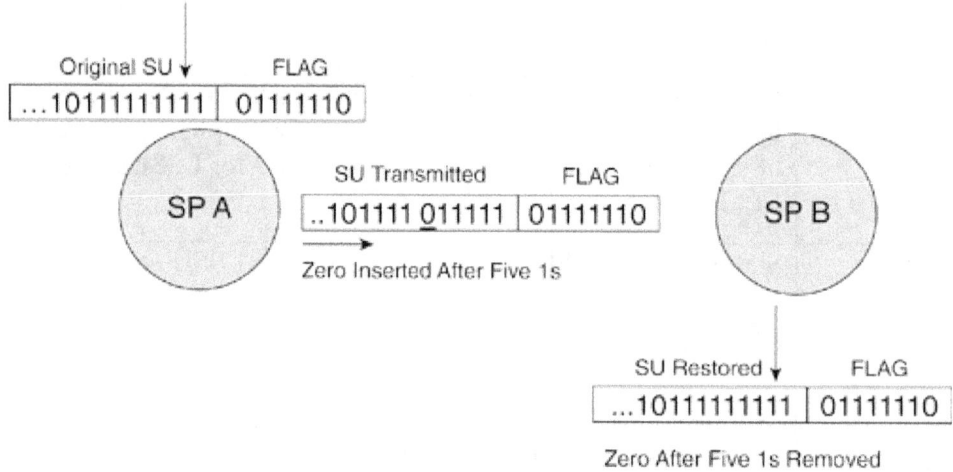

Figure 5-5: Zero Bit Insertion and Deletion

As another example, if the pattern 01111100_{LSB} appears in the SU, the pattern is changed to 001111100_{LSB} and then is changed back at the receiving end.

5.2.2.6 Length Indicator

MTP2 must be able to determine the SU type to process it. The length indicator (LI) provides an easy way for MTP2 to recognize the SU type. The LI indicates the number of octets between the LI and the CRC fields. Using telecommunications conventions, MTP2 measures the size of SUs in octets. An octet is simply another term for a byte; all SUs have an integral number of octets.

5.2.2.7 Signal Unit Alignment

Loss of alignment occurs when a nonpermissible bit pattern is received or when an SU greater than the maximum SU size is received.

MTP2 constantly processes the data stream, searching for flags that delineate the SUs. The maximum number of consecutive 1s that should be found in the bit stream is six (as part of the flag), because the transmitting end performs 0 bit insertion. If seven or more consecutive 1s are detected, this signifies a loss of alignment.

5.2.2.8 Error Detection

The error detection method is performed by a 16-bit CRC on each signal unit. These 16 bits are called check bits (CK bits).

NOTE

The process uses the Recommendation V.41 [ITU-T Recommendation V.41: CODE-INDEPENDENT ERROR-CONTROL SYSTEM, November 1988] generator polynomial $X^{16} + X^{12} + X^5 + 1$. The transmitter's 16-bit remainder value is initialized to all 1s before a signal unit is transmitted. The transmission's binary value is multiplied by X^{16} and then divided by the generator polynomial. Integer quotient values are ignored, and the transmitter sends the complement of the resulting remainder value, high-order bit first, as the CRC field. At the receiver, the initial remainder is preset to all 1s, and the same process is applied to the serial incoming bits. In the absence of transmission errors, the final remainder is 1111000010111000 ($X^0 + X^{15}$). [2]

5.2.2.9 Error Correction

Two methods of error correction are available: basic error correction (BEC) and preventive cyclic retransmission (PCR) method. The basic method is used for signaling links using non intercontinental terrestrial transmission and for intercontinental links that have a one-way propagation delay of less than 30 ms. The PCR method is used for all signaling links that have a propagation delay greater than or equal to 125 ms and on all satellite signaling links. Where the one-way propagation delay is between 30 and 125 ms, other criteria must be considered that are outside the scope of this book. Depending on other additional criteria, PCR can also be employed on links that have a one-way propagation delay between 30 ms and 125 ms.

5.2.2.10 Basic Error Correction

Basic error correction is accomplished using a backwards retransmission mechanism, in which the sender retransmits the corrupt (or missing) MSU and all subsequent MSUs. This method uses both negative and positive acknowledgments. Positive acknowledgments (ACKs) indicate the correct reception of an MSU, and negative acknowledgments (NACKs) are used as explicit requests for retransmission..

The basic error correction fields occupy a total of two octets in each SU and consist of an FSN, BSN, FIB, and BIB. The Forward Sequence Number (FSN) and Backward Sequence Number (BSN) are cyclic binary counts in the range 0 to 127.

Sequence Numbering

Each SU carries two sequence numbers for the purpose of SU acknowledgment and sequence control. Whereas the FSN is used for the function of SU sequence control, the BSN is used for the function of SU acknowledgment. [2]

Before it is transmitted, each MSU is assigned an FSN. The FSN is increased linearly as MSUs are transmitted. The FSN value uniquely identifies the MSU until the receiving SP accepts its delivery without errors and in the correct sequence.

Positive Acknowledgment

When the BIB in the received SU has the same value as the FIB that was sent previously, this indicates a positive acknowledgment.

The receiving SP acknowledges positive acceptance of one or more MSUs by copying the FSN value of the last accepted MSU into the SU's BSN, which it transmits..

Negative Acknowledgment

When the BIB in the received SU is not the same value as the FIB that was sent previously, this indicates a negative acknowledgment.

The receiving SP generates a negative acknowledgment for one or more MSUs by toggling the BIB's value. It then copies the FSN value of the last accepted MSU into the SU's BSN, which it transmits in the opposite direction.

5.2.2.10.1 Examples of Error Correction

Figure 5-6 shows the fundamental principles of basic error correction by examining the error correction procedure for one direction of transmission between SP A and SP B. A similar relationship exists between the FSN/FIB from SP B and the BSN/BIB from SP A.

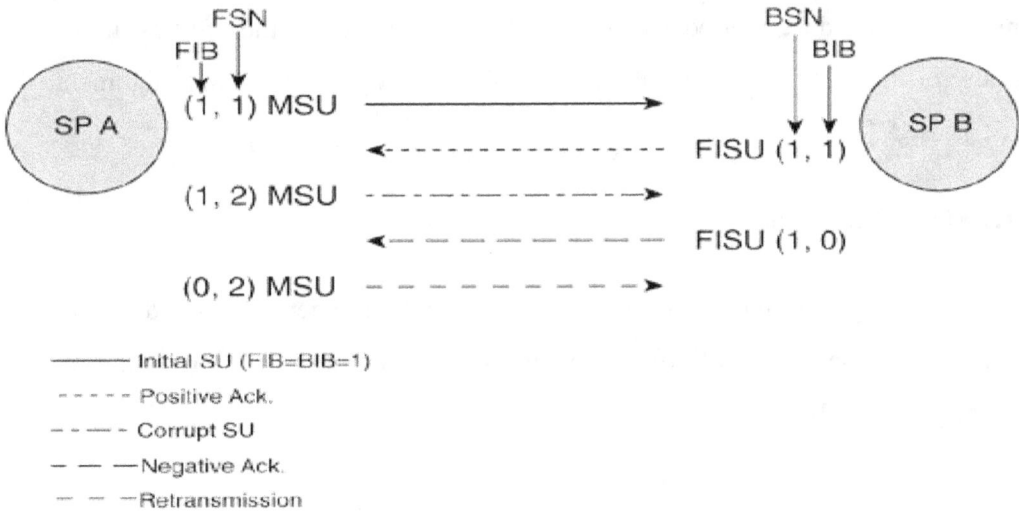

Figure 5-6: Principles of Basic Error Correction.

Although the bit rate on the signaling link in either direction is the same, note that the number of SUs transmitted by the two SPs in a time interval is likely to differ because of the MSUs' variable lengths. As a consequence, an SP might receive a number of SUs before it can acknowledge them.

5.2.2.10.2 Comparison with the Basic Error Correction Method

The basic error correction method is preferred on links that have one-way propagation times of less than 30 ms, because this allows higher MSU loads than with PCR. PCR achieves lower MSU loads because it expends a relatively large amount of time needlessly retransmitting MSUs that have already been received correctly (even though they have not yet been acknowledged). PCR links are highly underutilized because spare capacity is required to ensure that retransmissions can take place.

5.2.2.11 Signaling Link Initial Alignment

Initial alignment is performed for both initial activation of the link (power on) to bring it to service and to restore a link following a failure. Alignment is based on the compelled exchange of status information and a proving period to ensure that SUs are framed correctly. MTP3 requests initial alignment, which is performed by MTP2. Because MTP2 operates independently on each link, the initial alignment procedure is performed on a single link without involving other links. There are two forms of alignment procedures: the emergency procedure and the normal alignment procedure. The emergency procedure is used when the link being aligned is the only available link for any of the routes defined within the SSP. Otherwise, the normal alignment procedure is used. [2]

5.2.2.11.1 Status Indications

LSSUs are exchanged as part of the alignment procedure. There are six different status indications. The alignment procedure passes through a number of states during the initial alignment: [2]

- Idle
- Not Aligned
- Aligned

- Proving
- Aligned/Ready
- In Service

Idle

When an SP is powered up, the links are initially put in the idle state. The idle state is the first state entered in the alignment procedure; it indicates that the procedure is suspended. If the procedure fails at any time, it returns to the idle state.

Not Aligned

When MTP2 receives an order to begin initial alignment, the SP changes the status of the transmitted LSSUs to indication SIO (out of alignment) and starts the timer T2. If T2 expires, the status of the transmitted LSSUs reverts to SIOS.

Aligned

During T2 SIO, if SIN (normal alignment) or SIE (emergency alignment) is received from the remote SP, T2 is stopped, and the transmission of SIO ceases. The SP then transmits SIN or SIE, depending on whether normal or emergency alignment has been selected and timer T3 is started.

Proving

The proving period is shorter for emergency alignment and as a result is not as thorough. As previously stated, emergency alignment is selected if only one in service (or none) exists between two SPs. If the local SP detects an emergency alignment situation, emergency alignment is used regardless of whether an SIN or SIE is received from the distant SP. Similarly, emergency alignment is used if an SIE is received from the distant SP, even when the local MTP3 indicates a normal alignment situation (more than one in-service link between the two adjacent nodes).

Aligned/Ready

When T4 expires, the transmission of SIN/SIE ceases, timer T1 is started, and FISUs are transmitted. If timer T1 expires, the transmission of FISUs ceases, and LSSUs of type SIOS are transmitted.

In Service

Timer T1 stops upon receiving either FISUs or MSUs. When it stops, the SUERM becomes active. Figure 5-7 shows the initial alignment procedure.

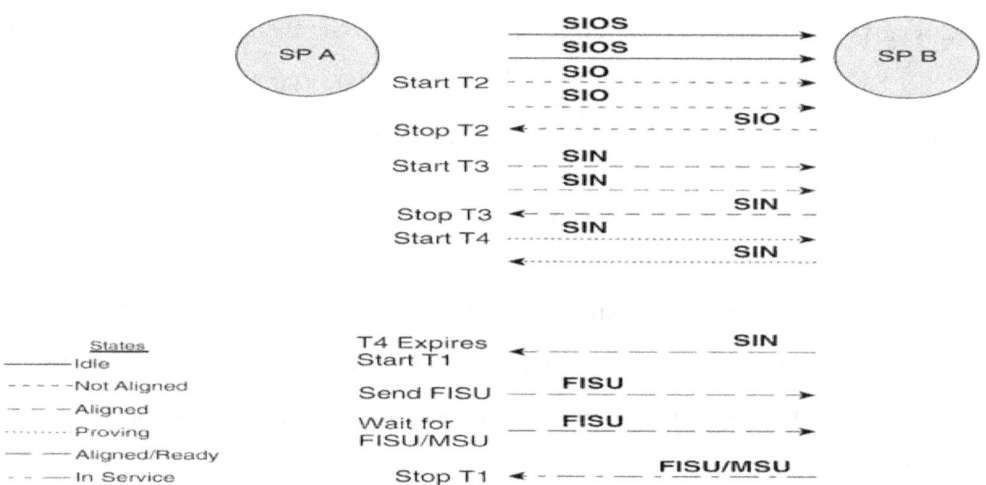

Figure 5-7: Procedure for Signaling Link Alignment

5.2.2.12 Signaling Link Error Monitoring

Error rate monitoring is performed both for an in-service link and when the initial alignment procedure is performed. Signal Unit Error Rate Monitor (SUERM) and the Alignment Error Rate Monitor (AERM) are the two link error rate monitors that are used. The SUERM performs monitoring when the link is in service, and the AERM performs monitoring when the link is undergoing initial alignment to bring it into service. The following sections describe these two link error rate monitors. [2]

5.2.2.12.1 SUERM

The SUERM is active when a link is in service, and it ensures the removal of a link that has excessive errors. It employs a leaky bucket counter, which is initially set to 0. The counter is increased by 1 for each SU that is received in error. The counter is decreased by 1 for each block of D consecutive SUs received without error, if it is not at 0. If the link reaches a threshold of T, MTP2 informs MTP3, which removes it from service. For a 64-kbps link, the values of D and T are 256 and 64, respectively.

5.2.2.12.2 AERM

The AERM is active when the link is in the proving period of the initial alignment procedure. The counter is initialized to 0 at the start of the proving period and is increased for every LSSU that is received in error.

The values of the four parameters for 64-kbps and lower bit rates (both for ITU and ANSI) are as follows:

- $T_{in} = 4$
- $T_{ie} = 1$
- $M = 5$
- $N = 16$

5.2.2.13 Flow Control

Flow control allows incoming traffic to be throttled when the MTP2 receive buffer becomes congested. When an SP detects that the number of received MSUs in its input buffer exceeds a particular value—for example, because MTP3 has fallen behind in processing these MSUs—it begins sending out LSSUs with the status indicator set to busy (SIB). These LSSUs are transmitted at an interval set by timer T5, sending SIB (80 to 120 ms), until the congestion abates. The congested SP continues sending outgoing MSUs and FISUs but discards incoming MSUs. Timer T6, remote congestion, is started when the initial SIB is received. If timer T6 expires, it is considered a fault, and the link is removed from service. [2]

Figure 5-8 depicts flow control using LSSUs with status indication busy.

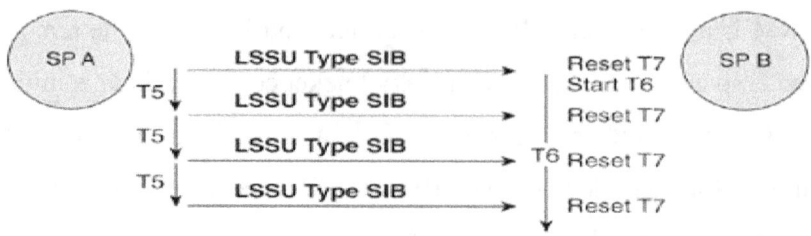

Figure 5-8: Flow Control Using Status Indication SIB

5.2.3 MTP 3 Functions

5.2.3.1 Message Format

The MTP3 portion of an SS7 message consists of two fields: the Signaling Information Field (SIF) and the Service Information Octet (SIO). The SIF contains routing information and the actual payload data being transported by the MTP3 service. The SIO contains general message characteristics for identifying the network type, prioritizing messages (ANSI only), and delivering them to the appropriate MTP3 user. When an SS7 node receives messages, Signaling Message Handling (SMH) uses the SIO and the portion of the SIF that contains routing information to perform discrimination, routing, and distribution. [2]

5.2.3.2 ITU-T Routing Label

The ITU-T routing label consists of the following fields:

- DPC
- OPC
- SLS

The ITU-T point codes are 14 bits in length. For ITU-T national networks, all 14 bits are interpreted as a single identifier that is often referred to as a structureless Point Code. For international networks, an International Signaling Point Code (ISPC) is subdivided into hierarchical fields, shown in Figure 5-9

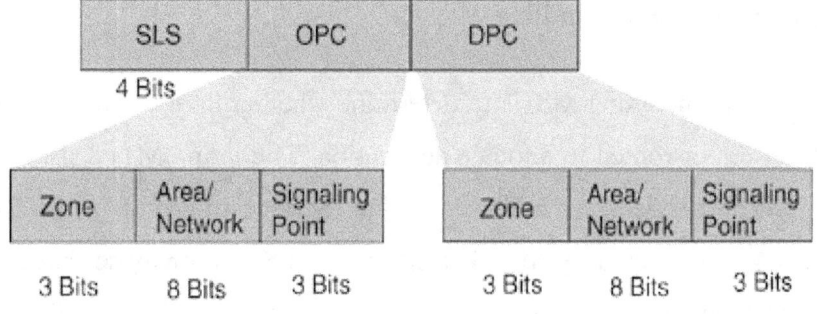

Figure 5-9: ITU-T Routing Label

5.2.3.3 ANSI Routing Label

The ANSI routing label consists of the following fields:

- DPC
- OPC
- SLS

The ANSI Point Code is 24 bits in length and is subdivided into three fields of one octet each, as shown in Figure 5-10 The three octets define the network, cluster, and member that uniquely identify the signaling node within the network hierarchy. The SLS field is an eight-bit field used for selecting the link and/or linkset for message transmission. This field was only five bits in earlier versions of the protocol, but was extended for better load sharing across signaling links in the 1996 version of the ANSI standards.

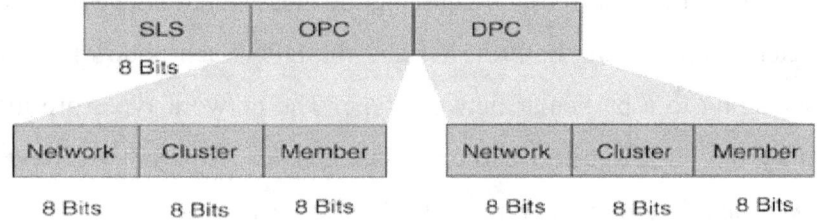

Figure 5-10: ANSI Routing Label

5.2.3.4 Signaling Message Handling

MTP3 processes all incoming MSUs to determine whether they should be sent to one of the MTP3 users or routed to another destination. The term "MTP3 user" refers to any user of MTP3 services, as indicated by the Service Indicator in the SIO. This includes messages generated by MTP3 itself, such as SNM, or those that are passed down from the User Parts at level 4 of the SS7 protocol, like ISUP and SCCP. The term "MTP User Part" is also used, but more specifically refers to the User Parts at level 4. When a node generates an MSU, MTP3 is responsible for determining how to route the message toward its destination using the DPC in the Routing Label and the Network Indicator in the SIO. Figure 5-11 shows how MTP3 message processing can be divided into three discrete functions: discrimination, distribution, and routing.

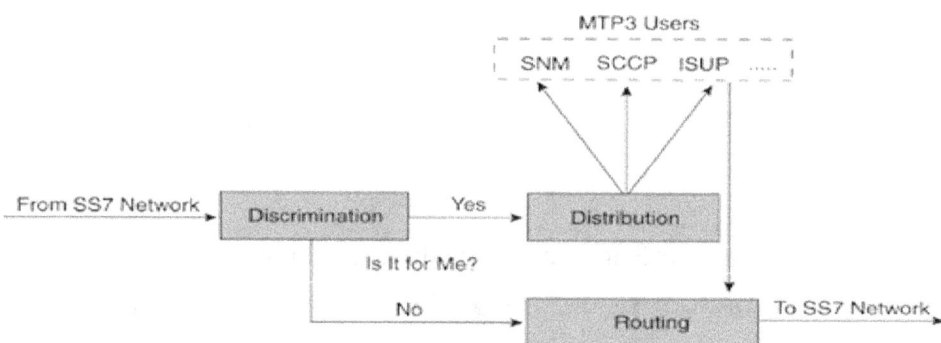

Figure 5-11: Signaling Message Handling

Discrimination

Message discrimination is the task of determining whether an incoming message is destined for the node that is currently processing the message. Message discrimination makes this determination using both the NI and the DPC. Each node's Point Code is defined as belonging to a particular network type. The network types are those that are specified by the NI, discussed earlier in this chapter. An ISC will have both a National network and International type, with Point Codes in each.

Each time a node receives a message, it must ask, "Is it for me?" The node asks the question by comparing the incoming DPC in the Routing Label to its own Point Code. If the Point Codes match, the message is sent to Message Distribution for processing.

If the Point Codes do not match, the message is sent to the Routing function if the node is capable of routing.

Distribution

When the discrimination function has determined that a message is destined for the current node, it performs the distribution process by examining the Service Indicator, which is part of the SIO in the Routing Label. The Service Indicator designates which MTP3 user to send the message to for further processing.

Routing

Routing takes place when it has been determined that a message is to be sent to another node. There are two circumstances in which this occurs. The first is when a node originates a message to be sent to the network. For example, an MTP3 user (such as ISUP or SCCP) generates a message for MTP3 to send. The second is when an STP has received a message that is destined for another node. The routing function is invoked if the discrimination function has determined that the received message is not destined for the STP. If a Signaling End Point (SSP or SCP) receives a message and the discrimination function determines that the message is not for that node, the message is discarded because these nodes do not have transfer capability. A User Part Unavailable (UPU) is sent to the originating node to indicate that the message could not be delivered. In other words, SEPs can only route the messages they originate. A node examines one or more routing tables to attempt to find a match for the DPC to which the message is to be routed. [2]

5.2.3.5 Message Load Sharing

A properly designed SS7 network employs alternate message paths to create network redundancy. User traffic is typically load-shared across different paths to maintain a balanced load on network equipment. Load sharing also ensures that problems on each path are detected quickly because they are carrying traffic. There are two types of SS7 load sharing:

- Load-sharing across linksets in a combined linkset
- Load-sharing across links within a linkset

When load sharing is used, the SLS field determines the distribution of messages across linksets and links as they traverse the network. The originating node generates an SLS code and places it into the Routing Label. At each node in the message path the SLS is used to map the message to a specific link and, if using a combined linkset, to a specific linkset.

5.2.3.6 Comparing the IP and MTP3 Protocols

The MTP3 message handling is similar to the Internet Protocol (IP) in some respects. For those who are familiar with IP, a comparison of the two protocols helps to put MTP3 in perspective. This is not intended to suggest an exact comparison; rather, to relate something that is known about one protocol to something similar in the other. The main point is that both protocols are packet based and designed to deliver messages to a higher layer service at a node in the network. It is not surprising that there are a number of commonalities given that the requirements are similar. In fact, studying a number of communications transport protocols shows that many share a common functionality and structure, with each diverging slightly to address its particular requirements. Table 5-3 lists an association of key IP packet fields with their MTP3 counterparts.

Table 5-3: Comparison of IP and MTP3 Packet Fields

IP	SS7
Source IP Address	Originating Point Code
Destination IP Address	Destination Point Code
Protocol	Service Indicator
Precedence (part of TOS field)	Priority
Data	User Data

5.2.3.7 Signaling Network Management

Failures in the SS7 network have potentially devastating effects on the communications infrastructure. The loss of all SS7 signaling capabilities at an SP isolates it from the rest of the network. The SS7 networks in existence today are known for their reliability, primarily due to the robustness of the SS7 protocol in the area of network management. Of course, this reliability must be accompanied by good network design to provide sufficient network capacity and redundancy. MTP3 Network Management is comprised of a set of messages and procedures that are used to ensure a healthy signaling transport infrastructure. This involves automatically invoking actions based on network events, such as link or route failures and reporting network status to other nodes.

Signaling Network Management is divided into three processes:

- Traffic management
- Route management
- Link management

Traffic management is responsible for dealing with signaling traffic, which are the messages generated by MTP3 users, such as ISUP and SCCP. The goal of Traffic management is to keep traffic moving toward its destination, even in the event of network failures and congestion, with as little message loss or mis-sequencing as possible.

Route management exchanges information about routing status between nodes. As events occur that affect route availability, route management sends messages to notify other nodes about the change in routing states..

Link management activates, deactivates, and restores signaling links. This involves notifying MTP users of the availability of signaling links and invoking procedures to restore service when a disruption has occurred.

5.2.3.8 Link Management

Links are physical entities that are made available to MTP3 users when they have proven worthy of carrying messages. If a link fails, it has a direct impact on the two nodes the link connects. It is link management's responsibility to detect any communication loss and attempt to restore it. Both nodes connected to the link invoke procedures for restoration in an attempt to restore communication. Link management can be divided into three processes:

- Activation
- Deactivation
- Restoration

Activation is the process of making a link available to carry MTP3 user traffic. Maintenance personnel typically perform it by invoking commands from an OAM interface to request that the link be activated for use. [2]

Deactivation removes a link from service, making it unavailable for carrying traffic. Like activation, this process is typically initiated by invoking commands from an OAM interface.

Restoration is an automated attempt to restore the link to service after a failure, making it available for traffic management use. The link alignment procedure is initiated when level 2 has detected a link failure.

5.2.3.9 Route Management

Signaling route management communicates the availability of routes between SS7 nodes. Failures such as the loss of a linkset affect the ability to route messages to their intended destination. A failure can also affect more than just locally connected nodes. For example, the linkset between STP1 and SSP B has failed in Figure 5-12 As a result, SSP A should only route messages to SSP B through STP1 as a last resort because STP1 no longer has an associated route. Even though none of the links belonging to SSP A have failed, its ability to route messages to SSP B is affected. Signaling route management provides the means to communicate these types of changes in route availability using Signaling Network Management messages. [2]

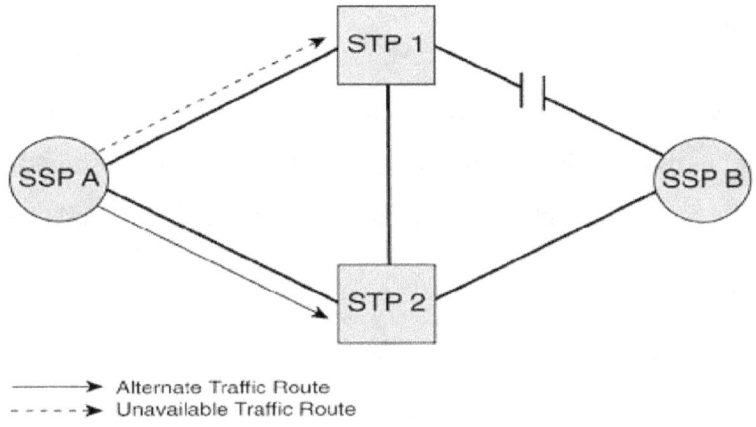

Figure 5-12: How Loss of Linkset Affects Routes

Route management uses the following messages to convey routing status to other network nodes:

- Transfer Prohibited (TFP)
- Transfer Restricted (TFR)
- Transfer Allowed (TFA)
- Transfer Controlled (TFC)

The following additional messages are used for conveying the routing status of clusters. They are only used in ANSI networks:

- Transfer Cluster Prohibited (TCP)
- Transfer Cluster Restricted (TCR)

Each node maintains a state for every destination route. As route management messages are received, the state is updated based on the status conveyed by the message. This allows nodes to make appropriate routing choices when sending messages. Routes can have one of three different states:

- Allowed
- Prohibited
- Restricted

Transfer Restricted

The restricted state indicates a limited ability to route messages. This status signifies that the primary route is unavailable and that another route should be chosen, if it exists. If the restricted route is the last available route in a routeset, it is still used for routing.

Transfer Prohibited

The Transfer Prohibited state indicates a complete inability to route messages to the affected destination. If one exists, another route must be chosen for routing. If no route exists, traffic management is notified that it cannot route messages to the destination.

Transfer Allowed

The transfer allowed state indicates that a route is available for carrying traffic. This is the normal state for in-service routes. When a route has been in the restricted or prohibited state and full routing capability is restored, the route's status is returned to transfer allowed.

Transfer Controlled

The Transfer Controlled message is used to indicate congestion for a route to a particular destination. The TFC message implies "transmit" congestion, in contrast to the "receive" buffer congestion handled by MTP2.

5.2.3.10 Traffic Management

Traffic management is the nucleus of the MTP network management layer that coordinates between the MTP users' communication needs and the available routing resources. It is somewhat of a traffic cop in stopping, starting, redirecting, and throttling traffic. Traffic is diverted away from unavailable links and linksets, stopped in the case of unavailable routesets, and reduced where congestion exists.

Traffic management depends on the information provided by link management and route management to direct user traffic. For example, when a TFP is received for a destination, traffic management must determine whether an alternate route is available and shift traffic to this alternate route. During this action, it determines what messages the unavailable destination has not acknowledged so those messages can be retransmitted on the alternate route. This section discusses the following procedures that are employed by traffic management to accomplish such tasks: [2]

- Changeover
- Emergency changeover
- Time-controlled changeover
- Changeback
- Time-controlled diversion
- Forced rerouting
- Controlled rerouting
- MTP restart
- Management inhibiting

5.15 Summary

MTP2 provides monitoring functions to MTP3 by using error rate counters. If specified thresholds are exceeded, MTP3 asks MTP2 to put the link out of service. If instructed by MTP3 to do so, MTP2 attempts to put specified links in service by following an alignment procedure. MTP2 also provides status indications when it encounters congestion and when layers above MTP2 can no longer process MSUs because of failure.

MTP3 adheres to the modularity of the OSI model, the user parts can depend on the MTP3 transport without being aware of the underlying details. The two levels exchange a simple set of primitives to communicate status.

Chapter 6

Signaling Connection Control Part (SCCP)

6.1 Introduction

Signaling Connection Control Part (SCCP), a routing protocol in SS7 protocol suite in layer 4, provides end-to-end routing for TCAP messages to their proper database. The SCCP provides additional network layer functions to provide transfer of noncircuit-related (NCR) signaling information, application management procedures and alternative and more flexible methods of routing. SCCP relies on the services of MTP for basic routing and error detection. [17]

The Signaling Connection Control Part (SCCP) layer of the SS7 stack provides connectionless and connection-oriented network services and global title translation (GTT) capabilities above MTP Level 3. SCCP is used as the transport layer for TCAP-based services. It offers both Class 0 (Basic) and Class 1 (Sequenced) connectionless services. SCCP also provides Class 2 (connection oriented) services, which are typically used by Base Station System Application Part, Location Services Extension (BSSAP-LE). In addition, SCCP provides Global Title Translation (GTT) functionality.

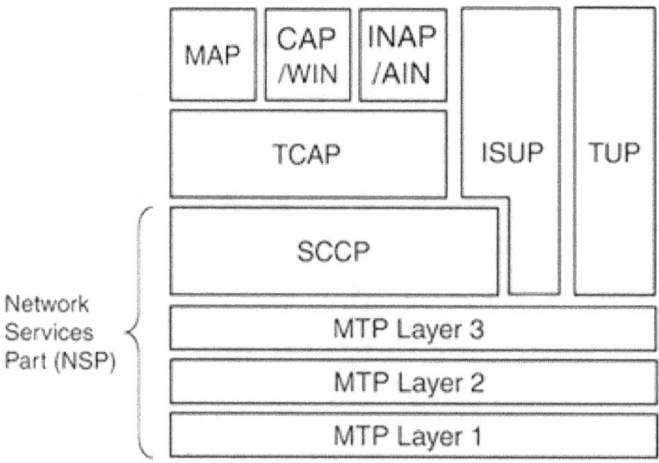

Figure 6-1: SS7 Stack with the Network Service Part (NSP) Highlighted

SCCP is used extensively in cellular networks. Base Station Subsystem Mobile Application Part (BSSMAP) and Direct Transfer Application Part (DTAP) use it to transfer radio-related messages in Global System for Mobile communication (GSM). In conjunction with Transfer Capabilities Application Part (TCAP), SCCP is also used throughout the GSM Network Switching Subsystem (NSS) to transport Mobile Application Part (MAP) signaling between the core GSM components to enable subscriber mobility and text messaging (SMS), among other items. For example, when the Visitor Location Register (VLR) queries the Home Location Register (HLR) to obtain the subscriber's profile, SCCP is responsible for transferring both the query and the response back to the VLR.

Fixed-line networks primarily use SCCP for intelligent network applications and advanced supplementary services. Fixed-line intelligent networks use Advanced Intelligent Network (AIN) within North America and Intelligent Network Application Protocol (INAP) outside of North America. AIN/INAP both use SCCP's transport, application management, and enhanced routing functionalities. Two example supplementary services that require the use of SCCP include CCBS and Completion of Calls on No Reply (CCNR). [2]

6.2 SCCP Architecture

SCCP is composed of the following four functional areas:

- ✓ SCCP connection-oriented control (SCOC) —Responsible for setting up and releasing a virtual connection between two SCCP users. SCOC can offer features including sequencing, flow control, and segmentation and can override congestion procedures by assigning data priority.
- ✓ SCCP connectionless control (SCLC) —Responsible for transferring data between SCCP users without creating a virtual connection. In addition to segmentation, it can perform limited sequencing.
- ✓ SCCP routing control (SCRC) — Provides additional routing beyond that offered by MTP3, through the use of global titles.
- ✓ SCCP management (SCMG) — Responsible for tracking application status and informing SCMG at other SCCP nodes, as necessary.

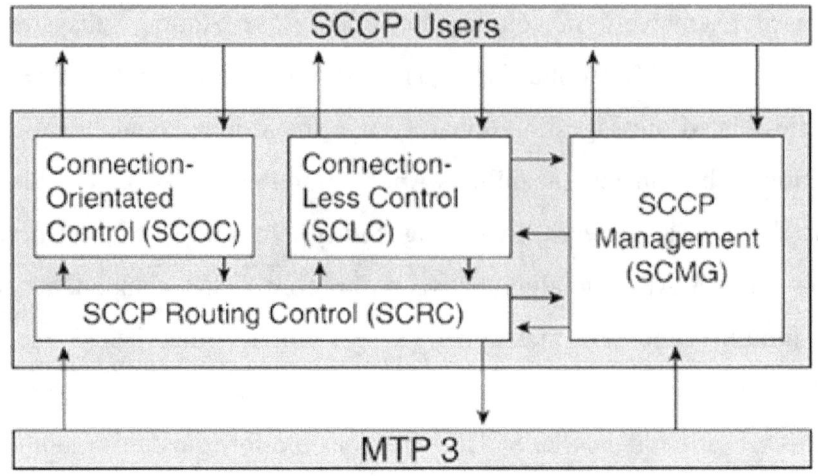

Figure 6-2: The SCCP Architecture

6.2.1 SCCP Connectionless Control (SCLC)

SCLC is used to provide the capabilities that are necessary to transfer one NSDU in the "data" field of a UDT, Long Unit Data (LUDT), and XUDT message. The SCLC routes the message without regard to the route that the messages follow through the network. These services are provided without setting up a logical connection.

SCLC formats the user data into a message of the appropriate protocol class (0 or 1 in the case of connectionless) and transfers it to SCRC for routing. On receiving a message, SCLC is responsible for decoding and distributing the message to the appropriate subsystem. Figure 6-3 shows data transfer using SCLC: data is simply sent without the prior establishment of references at each side.

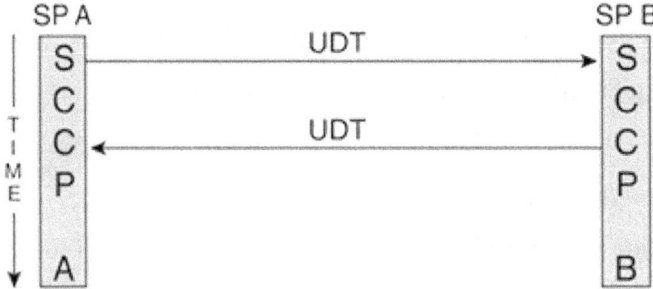

Figure 6-3: The Transfer of Connectionless Messages from One SCCP User to Another

6.2.2 SCCP Connection-Oriented Control (SCOC)

SCOC is used to route messages through a specific, fixed logical network path. To establish a dedicated logical connection between an originating SCCP user (subsystem) and a terminating SCCP user (subsystem), the SCCP users residing at different nodes throughout the network communicate with each other.

A signaling connection between the SCCP users is established, making both SCCP users aware of the transaction by using the DLR and SLR parameters. The signaling connection is released at the end of the transaction (information transfer). This is similar to SS7 protocol TUP/ISUP, which is used to control telephony calls, in that a connection is setup and released at a later time. However, the connection is virtual; there is not a trunk with user traffic being set up and released—rather, there is a virtual connection over the signaling network for the purpose of data transfer between applications (subsystems).

Connection-oriented procedures can be split into three phases:

- ✓ Connection Establishment Phase— The SCCP users set up a logical, fixed path that the data packets will follow. The path might involve only two or three nodes with SCCP capability or, depending on how many intermediate nodes exist between the originator and terminator, it might involve a much larger number.
- ✓ Data Transfer Phase— After the connection is established, the data that is to be transferred is converted into an NSDU and sent in a DT1 or DT2 message. Each NSDU is uniquely identified as belonging to a specific signaling connection. In this way, it is possible for the SCCP to simultaneously handle independent signaling connections.
- ✓ Connection Release Phase— After all NSDUs have been transmitted and confirmed, either or both of the user applications that initiated the process release the logical path. A release can also occur if the connection fails.

An example of a connection-oriented data transfer is carried out in Figure 6-4 at the request of the SCCP user; SCCP A establishes a logical connection by sending a Connection Request (CR) message to SCCP B and assigning a SLR to the request. The remote node confirms the connection by sending a Connection Confirm (CC) message and includes its own SLR and a DLR that is equal to SCCP A's SLR. This gives both sides a reference for the connection.

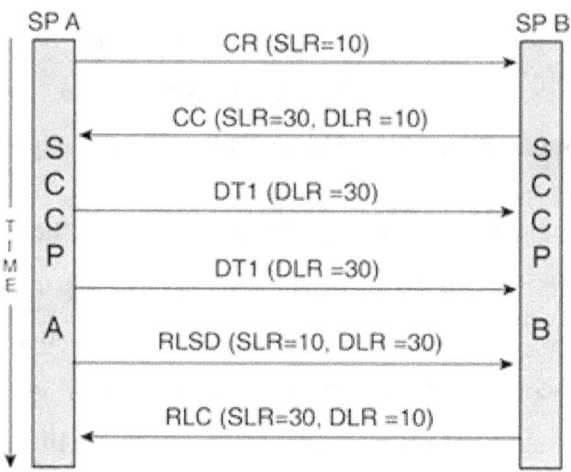

Figure 6-4: The Transfer of Connection-oriented Messages from One SCCP User to Another Using a Temporary Connection

The CR message contains the address of the destination SCCP node and user. The subsequent data message DT1 only needs to send the DLR because the logical connection has been established through the proceeding exchange of SLR and DLR. The clear-down messages contain both SLR and DLR. If intermediate nodes are involved, they make associations between pairs of SLR/DLRs to establish the logical connection. Upon release, the SLR/DLR references are available for further use on other transactions. SCCP nodes can establish multiple simultaneous logical connections through the use of the SLR and DLR.

6.2.3 SCCP Routing Control (SCRC)

SCRC performs the following three functions:

- Routes messages received from the MTP to appropriate local subsystem.
- Routes messages from local subsystems to other local subsystems.
- Routes messages from local subsystems to subsystems in remote nodes by utilizing MTP's transport services. The destination is specified in the called party (CdPA) address parameter, which is supplied by the subsystem. The

address can contain a combination of point code, system number, and global title.

There are three types of SCCP address:

- Signaling point code (SPC)
- Subsystem number (SSN)
- Global title (GT)

SPC refers to the address of MTP. It is valid in a SS7 network only.

The MTP designated by SPC identifies a destination SP with the received DPC and performs routing to the destination SP. In addition, it identifies the user of the destination SP according to Service Indicator (SI).

SSN is the local addressing information used by SCCP to identify SCCP users in the same node.

GT applies when the destination network address is unknown to the originating node.

GT can be used to identify any signaling point and subsystem worldwide.

MTP, however, cannot perform routing according to GT. The SCCP must first translate the GT of the called party into DPC or DPC+SSN. In addition, when SCCP sends the DPC or DPC+SSN to MTP, it needs to specify the numbering plan of the GT.

The calling address and called address in a SCCP message may be any or the combination of SPC, SSN, and GT.

The MTP layer can only use point code routing. Using MTP point code routing, MSUs pass through the STPs until they reach the SP that has the correct DPC.

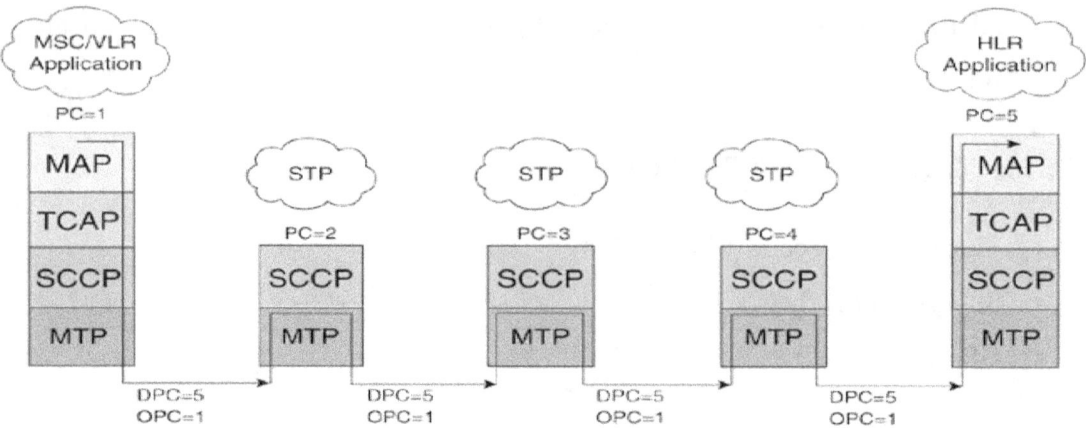

Figure 6-5: Showing MTP Point Code Routing

The following sections describe the SSN and GT routing.

6.2.3.1 Subsystem Number (SSN) Routing

As previously mentioned, a subsystem is the name given to an application that uses SCCP; applications are predominantly database driven, except where ISUP is the subsystem (for a limited number of supplementary services), or where BSSAP uses SCCP (for radio-related signaling in GSM).

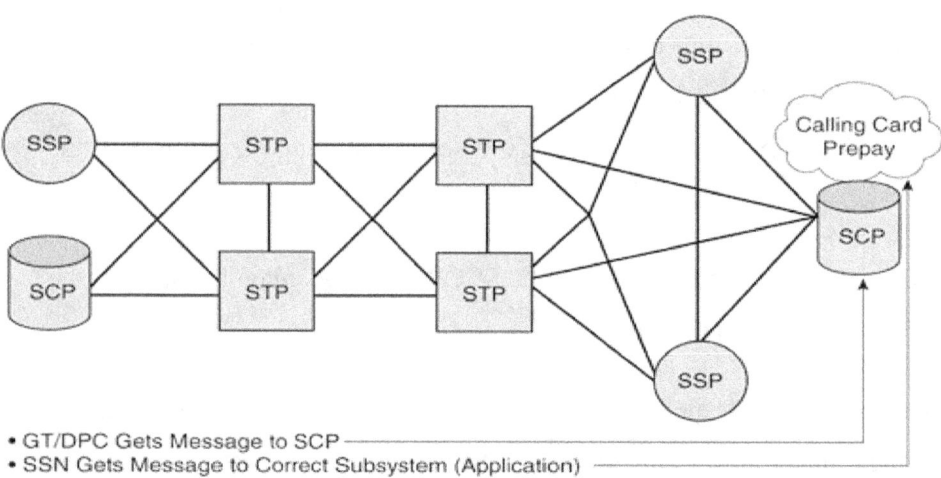

Figure 6-6: An SSN and DPC Are Required for the Final Delivery of an SCCP Message

Figure shows that a DPC and SSN are required in order to deliver a message to the correct application at the destination node.

6.2.3.2 Global Title (GT) Routing

"A global title is an address, such as dialed-digits, which does not explicitly contain information that would allow routing in the SS7 network."

There are many examples of digit strings that are global titles: in fixed-line networks, toll free, premium rate, numbers ported under LNP, or in the case of GSM cellular networks, the Mobile Subscriber ISDN Number (MSISDN) and International Mobile Subscriber Identity (IMSI) of the cellular subscriber and each HLR and VLR.

A GT is a telephony address. As such, the GT address must be translated into an SS7 network address (DPC+SSN) before it can be finally delivered. The GT is placed in the global title address information (GTAI) parameter within the CgPA and CdPA fields.

Global title routing is often used in fixed-line networks for calling-card validation and such services as telemarketing numbers (like a toll-free or premium rate). It is used in cellular networks for exchanging messages when an HLR and VLR belong to different networks or when several signaling points separate them.

Global Title Translation

GTT is an incremental indirect routing method that is used to free originating signaling points from the burden of having to know every potential application destination (that is, PC+SSN). This section describes the GTT process and the parameters associated with GTT.

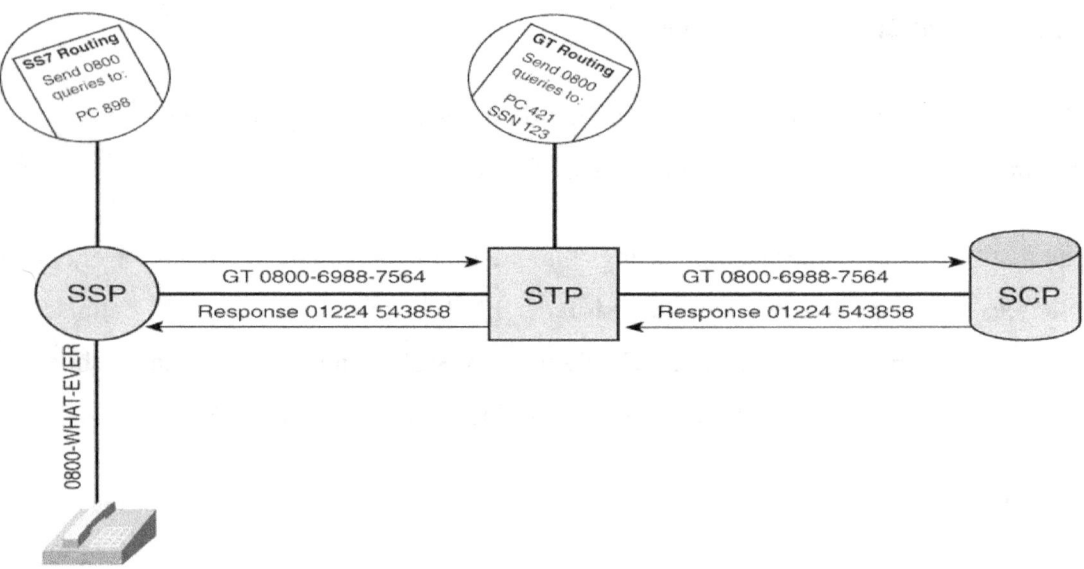

Figure 6-7: Example of GTT

6.3 SCCP Message Transfer Services

The SCCP provides two categories of service for data transfer: connection-oriented services and connectionless services. Within each service category, two classes of service are defined as follows:

- Class 0— Basic connectionless class
- Class 1— In-sequence delivery connectionless class
- Class 2— Basic connection-oriented class
- Class 3— Flow control connection-oriented class

6.3.1 Connectionless Services

Connectionless services are performed similar to the transfer of datagram in packet switching. No message transfer channel needs to be set up before message transfer. Signaling data are transferred on the signaling network. Therefore, routing function is provided by the SCCP.

In signaling message transfer, SCCP converts a called address into a signaling point code that can be recognized by MTP.

In connection services, messages are transferred as a whole (as unit data UDT) instead of being segmented.

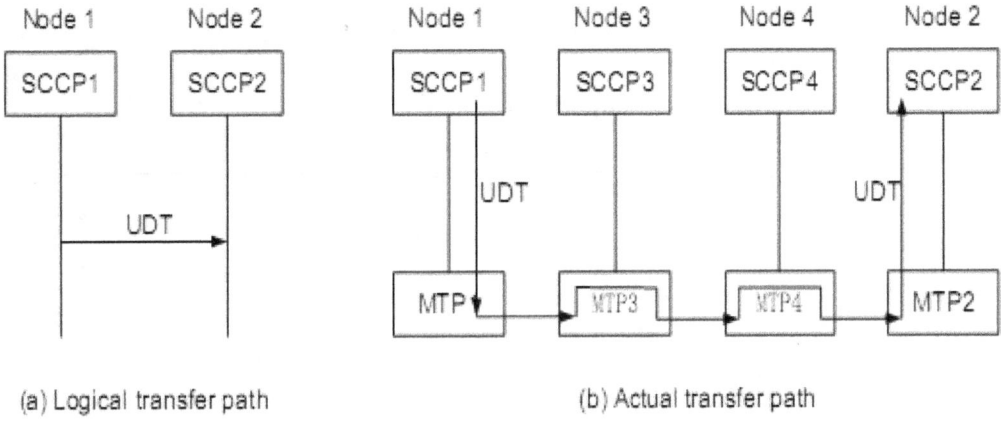

Figure 6-8: Connectionless transfer of signaling messages

Class 0 and Class 1 are connectionless services.

- ✓ Class 0: Basic connectionless service
  Signaling messages are transferred independent of one another. Therefore, there is no guaranteed in-sequence delivery of signaling messages to the destination signaling point.
- ✓ Class 1: In-sequence delivery connectionless service
  The data from the same information flow are attached with a signaling link selection code (SLS). Data packets with the same SLS are transferred on the same signaling link. Therefore, it is guaranteed that messages can be delivered to the destination signaling point in accordance with the transfer sequence.

6.3.2 Connection-Oriented Services

Connection-oriented services are performed similar to the packet switching over virtual circuits.

Before signaling message transmission, a message transfer channel (logical connection or virtual connection) is established between the originating node and the destination node.

The connection-oriented service is suitable for transferring a large amount of data.

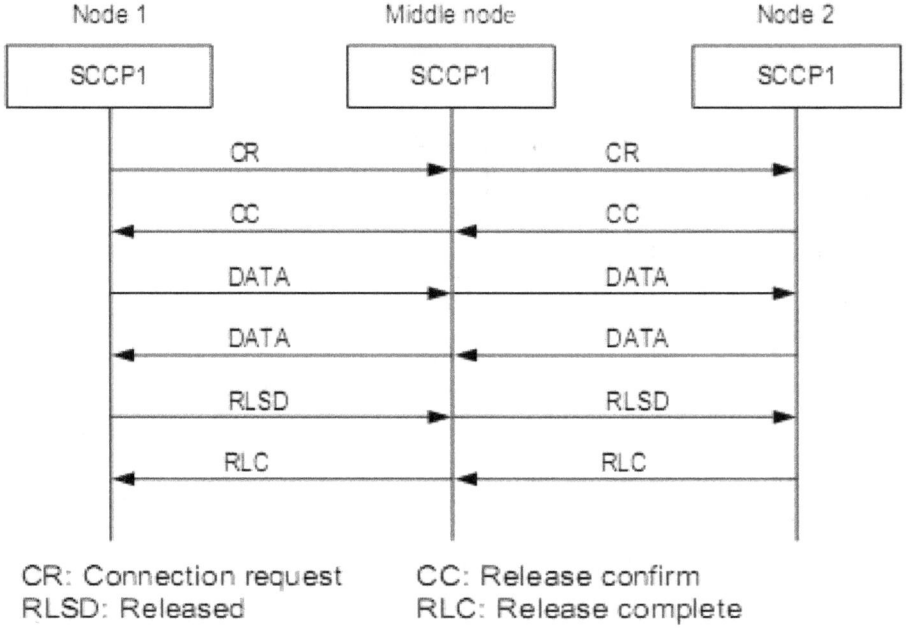

Figure 6-9: Connection-oriented transfer with a middle node

Class 2 and Class 3 are connection-oriented services

- ✓ Class 2: Basic connection-oriented class

    This type of service can guarantee that signaling messages are sent in the same sequence as they are received. Therefore, a long message can be transferred in segments, and reassembled after it is received.

- ✓ Class 3: Flow control connection-oriented class

    The features of class 2 are complemented by the inclusion of flow control, expedited data transfer, and detection of message loss or mis-sequencing.

    Connection-oriented services are classified into temporary signaling connections and permanent signaling connections.

    Temporary signaling connections are always under control (for example, during establishment, data transfer, and release). They are released after the transfer is finished. It is similar to a call connection.

    Permanent signaling connections are similar to permanent virtual circuits in packet data switching. They are established0 and controlled by the operation.
    [16]

6.4 SCCP Messages

SCCP messages encapsulated in the MSUs in MTP before transfer. For a MSU, SCCP message is its signaling information field (SIF).

6.4.1 Message Structure

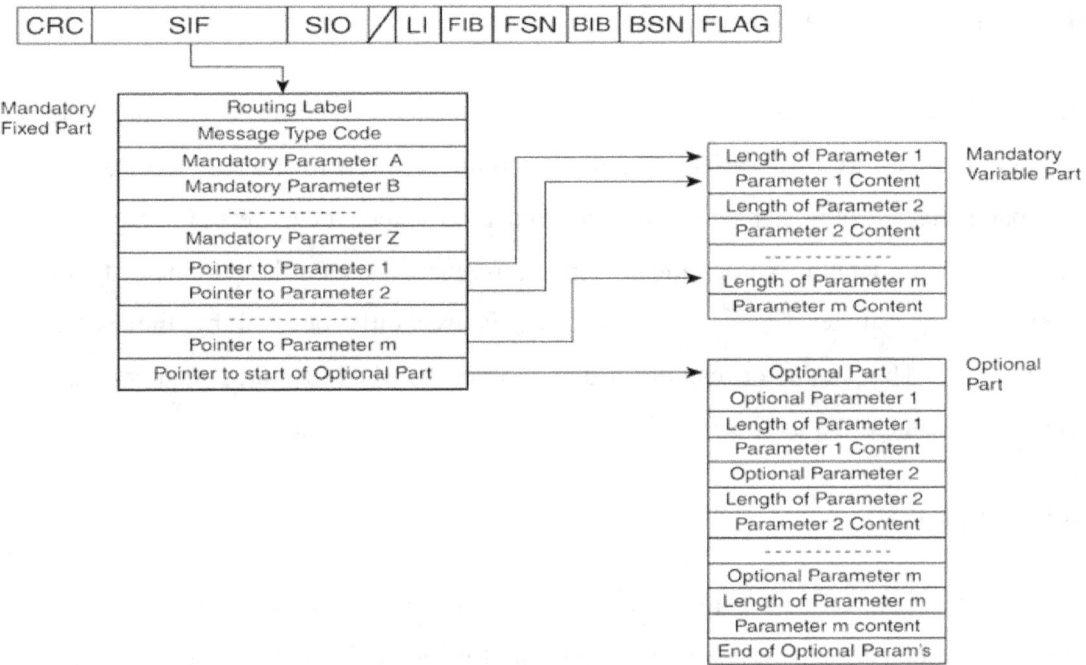

Figure 6-10: Structure of SCCP messages

With all other MTP users, SCCP messages are composed of three parts: a mandatory fixed part, mandatory variable part, and an optional part. All SCCP messages contain a mandatory fixed part, but not all of them have parameters to place in the mandatory variable or optional part. The following sections describe these three parts in more detail.

Mandatory Fixed Part (MF)

The mandatory fixed part consists of those parameters that must be present in the message and that are of a fixed length. Because the parameters are of a fixed length and are mandatory, no length indicator is required. In addition, because the parameter types and their order are known from the SCCP message type, no parameter names are required for stating the parameter types.

The mandatory fixed part contains pointers to the mandatory variable part and the optional part of the message. A pointer to the optional part is only included if the message type permits an optional part. If, on the other hand, the message type permits an optional part but no optional part is included for that particular message, then a pointer field that contains all zeros is used.

Mandatory Variable Part (MV)

The mandatory variable part consists of those parameters that must be present in the message and that are of a variable length. A pointer is used to indicate the start of each parameter. A length indicator precedes each parameter because the parameters are of a variable length. No parameter tags are required to state the parameter types because the parameter types and their order is explicitly defined by the SCCP message type. The parameters can occur in any order, but the associated pointers must occur in the same order as specified by the particular message type

Optional Part (O)

The optional part consists of those parameters that are not always necessary. Each parameter is preceded by a parameter name and a length indicator. The parameter name is a unique one-octet field pattern that is used to indicate the parameter type. Because the parameter types and their order are unknown, it is required for each parameter type.

A one-octet End of Optional Parameters field is placed at the end of the last optional parameter. It is simply coded as all zeros.

Figure 6-11 illustrates an example message that contains all three parts. The message could contain no optional parameters, or even more optional parameters than in the example shown. The following section details the CR message. [2]

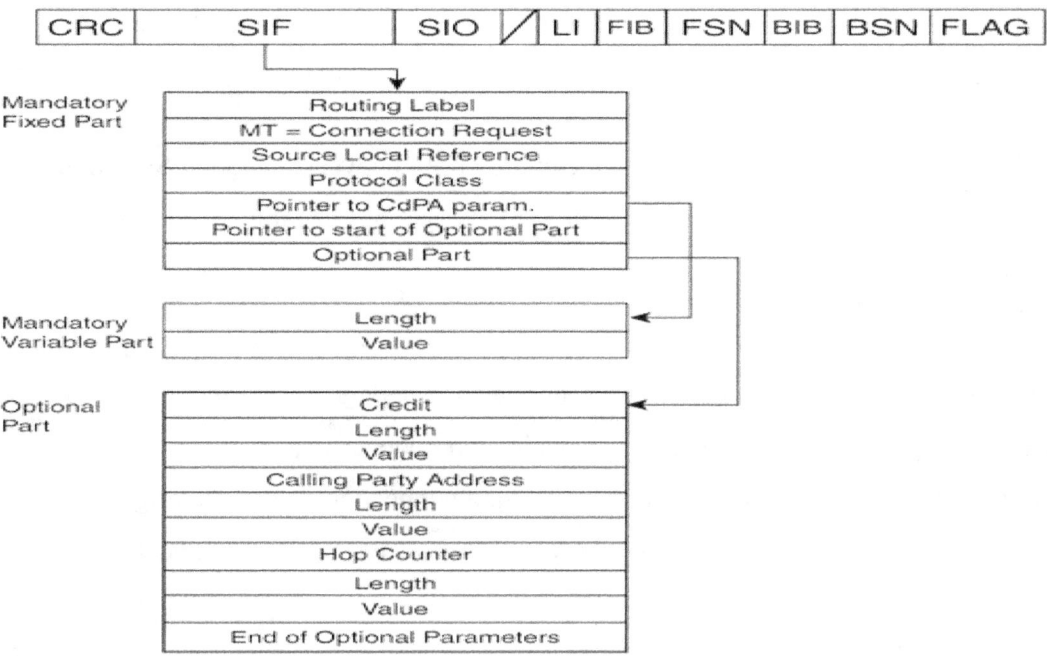

Figure 6-11: An Example of a Connection Request (CR) Message Structure

6.4.2 Message Types

Table shows the message types and their codes of various messages.

Table 6-1: SCCP message type and code

Message type	Protocol class				Message type code
	0	1	2	3	
Connection request (CR)			√	√	0000 0001
Connection confirm (CC)			√	√	0000 0010
Connection refused (CREF)			√	√	0000 0011
Released (RLSD)			√	√	0000 0100
Release complete (RLC)			√	√	0000 0101
Data form 1 (1DT1)			√		0000 0110
Data form 2 (2DT2)				√	0000 0111
Data acknowledgement (AK)				√	0000 1000
Unit data (UDT)	√	√			0000 1001
Unit data service (UDTS)	√	√			0000 1010
Protocol data unit error (ERR)			√	√	0000 1111
Inactivity test (IT)			√	√	0001 0000

Connection Request (CR)

Connection-oriented protocol Class 2 or 3 uses a CR message during the connection establishment phase. It is sent by an originating SCCP user to a destination SCCP user to set up a signaling connection (a virtual connection) between the two signaling points. The various parameters that compose the message dictate the connection requirements. After receiving the CR message, SCCP initiates the virtual connection setup, if possible.

In GSM cellular networks, a CR message could be used between a Mobile Switching Center (MSC) and a Base Station Controller (BSC) to setup a signaling connection.

Connection Confirm (CC)

Connection-oriented protocol Class 2 or 3 uses a CC message during the connection establishment phase. SCCP sends it at the destination node as an acknowledgement to the originating SCCP that it has set up the signaling connection. When the originating SCCP receives the CC message, it completes the setup of the signaling connection.

Connection Refused (CREF)

The connection-oriented protocol Class 2 or 3 can use a CREF message during the connection establishment phase. The destination SCCP or an intermediate node sends it to indicate to the originating SCCP that the signaling connection setup has been refused. As such, it is a negative response to a CR message. The refusal cause value is supplied to the originating SCCP.

In GSM cellular networks, a CREF message can be sent from an MSC to a BSC (or vice versa) to refuse the requested signaling connection because the SCCP of the signaling point (MSC or BSC) cannot provide the connection.

Released (RLSD)

The connection-oriented protocol Class 2 or Class 3 uses a RLSD message during the release phase. It is sent in the forward or backward direction to indicate that the sending SCCP wants to release the signaling connection.

In GSM cellular networks, a RLSD message is always sent from the MSC to the BSC (or vice versa) to release the SCCP connection and the resources that are associated with it.

Release Complete (RLC)

The connection-oriented protocol Class 2 or 3 uses a RLC message during the release phase. It is sent in the forward or backward direction as a response to the RLSD message to indicate the receipt of the RLSD and the execution of the appropriate actions for releasing the connection.

Data Form 1 (DT1)

Only connection-oriented protocol Class 2 uses a DT1 message during the data transfer phase. Either end of a signaling connection sends it to transparently pass SCCP user data between two SCCP nodes.

DT1 messages are used in cellular networks to transfer data between the BSC and MSC after CR and CC messages have established the connection. All data transfer between BSC and MSC is performed using DT1 messages. DT2 messages (used for protocol Class 3) are not used in GSM (or DCS1800).

Unitdata (UDT)

A UDT message is used to send data in connectionless mode using connectionless protocol Class 0 and Class 1.

UDT messages are commonly used for TCAP communication within IN services. In GSM cellular networks, UDT messages are used by the MAP protocol to send its messages.

Unitdata Service (UDTS)

A UDTS message is used in connectionless protocol Class 0 and 1. It indicates to the originating SCCP that a UDT message that is sent cannot be delivered to its destination. A UDTS message is only sent if the option field in the received UDT was set to return an error.

6.5 Summary

The SCCP provides additional OSI network layer functionality and, with the MTP, provides an NSP. It uses the signaling network to transport non circuit-related signaling, such as queries and responses between switches and telecommunications databases. SCCP provides two categories of service with two protocol classes in each. Classes 0 and 1 are within the connectionless category, and do not establish a virtual connection before transferring data. Classes 2 and 3 are within the connection-oriented category and establish a virtual (logical) connection before transferring data. SCCP provides flexible routing based on DPC, SSN, or GT, or a combination of all three. Global titles are an alias for a DPC and SSN and must be translated at nodes administered with the proper information (usually STPs).

Chapter 7

ISDN User Part (ISUP)

7.1 Introduction

The ISDN User Part (ISUP), a key protocol in the SS7/C7 signaling system, defines the protocol and procedures used to set-up, manage, and release trunk circuits that carry voice and data calls over the public switched telephone network (PSTN) between different switches. ISUP is used for both ISDN and non-ISDN calls. [18]

As its name implies, ISUP was created to provide core network signaling that is compatible with ISDN access signaling. The combination of ISDN access signaling and ISUP network signaling provides an end-to-end transport mechanism for signaling data between subscribers. ISUP provides signaling for both non-ISDN and ISDN traffic; in fact, the majority of ISUP-signaled traffic currently originates from analog access signaling, like that used by basic telephone service phones.

In addition to its speed efficiencies, ISUP enables more call-related information to be exchanged because it uses Common Channel Signaling (CCS). CAS signaling severely limits the amount of information that can be exchanged over trunks because it shares a small amount of space with a call's voice stream. [2]

When a telephone call is set up from one subscriber to another, many telephone exchanges will be involved, possibly across international boundaries. To allow a call to be set up correctly, where ISUP is supported, a switch will signal call-related information like called or calling party number to the next switch in the network using ISUP messages.

The features of ISUP are as follows: [16]

- Complete message types: Information carried in the message is abundant.
- Variable message length: Multiple parameters can be carried.
- Simple signaling program.
- Powerful functions: Supports various speech, non-speech, and supplementary services.

7.2 ISUP and the SS7 Protocol Stack

As shown in Figure 7-1, ISUP resides at Level 4 of the SS7 stack with its predecessor, the Telephone User Part (TUP). TUP is still used in many countries, but ISUP is supplanting it over time. TUP also provides a call setup and release that is similar to ISUP, but it has only a subset of the capabilities. TUP is not used in North America because its capabilities are not sufficient to support the more complex network requirements.

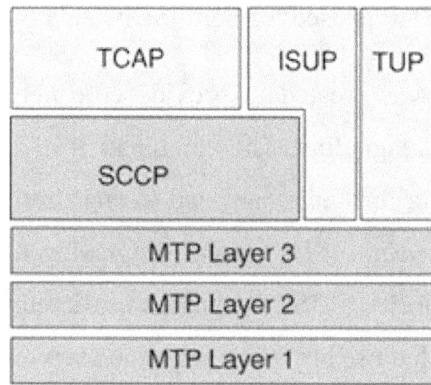

Figure 7-1: ISUP at Level 4 of the SS7 Stack

As you can see in Figure 7-1, a connection exists between ISUP and both the SCCP and MTP3 levels. ISUP uses the MTP3 transport services to exchange network messages, such as those used for call setup and clear down. The connection to SCCP is for the transport of end-to-end signaling. While SCCP provides this capability, today ISUP end-to-end signaling is usually transported directly over MTP3. The "Interworking with ISDN" section of this chapter further discusses end-to-end signaling and the two different methods using MTP3 and SCCP for transport.

ISUP supports basic bearer services and supplementary services of ISDN users, and realizes the functions of TUP and data user part (DUP). ISUP has more functions than TUP, but fewer message types. [2]

7.3 Basic ISUP Call Flow

Call set up

When a call is placed to an out-of-switch number, the originating SSP transmits an ISUP initial address message (IAM) to reserve an idle trunk circuit from the originating switch to the destination switch. The destination switch rings the called party line if the line is available and transmits an ISUP address complete message (ACM) to the originating switch to indicate that the remote end of the trunk circuit has been reserved. The STP routes the ACM to the originating switch which rings the calling party's line and connects it to the trunk to complete the voice circuit from the calling party to the called party.

Call connection

When the called party picks up the phone, the destination switch terminates the ringing tone and transmits an ISUP answer message (ANM) to the originating switch via its home STP. The STP routes the ANM to the originating switch which verifies that the calling party's line is connected to the reserved trunk and, if so, initiates billing.

Call tear down

If the calling party hangs-up first, the originating switch sends an ISUP release message (REL) to release the trunk circuit between the switches. The STP routes the REL to the destination switch. If the called party hangs up first, or if the line is busy, the destination switch sends an REL to the originating switch indicating the release cause (e.g., normal release or busy). Upon receiving the REL, the destination switch disconnects the trunk from the called party's line, sets the trunk state to idle, and transmits an ISUP release complete message (RLC) to the originating switch to acknowledge the release of the remote end of the trunk circuit. When the originating switch receives (or generates) the RLC, it terminates the billing cycle and sets the trunk state to idle in preparation for the next call. [19]

7.4 ISUP Message Flow

This section deals with an introduction to the core set of ISUP messages that are used to set up and release a call. The ISUP protocol defines a large set of procedures and messages, many of which are used for supplementary services and maintenance procedures.

A basic call can be divided into three distinct phases:

- Setup
- Conversation (or data exchange for voice-band data calls)
- Release

In Figure 7-2, part A illustrates the ISUP message flow for a basic call.

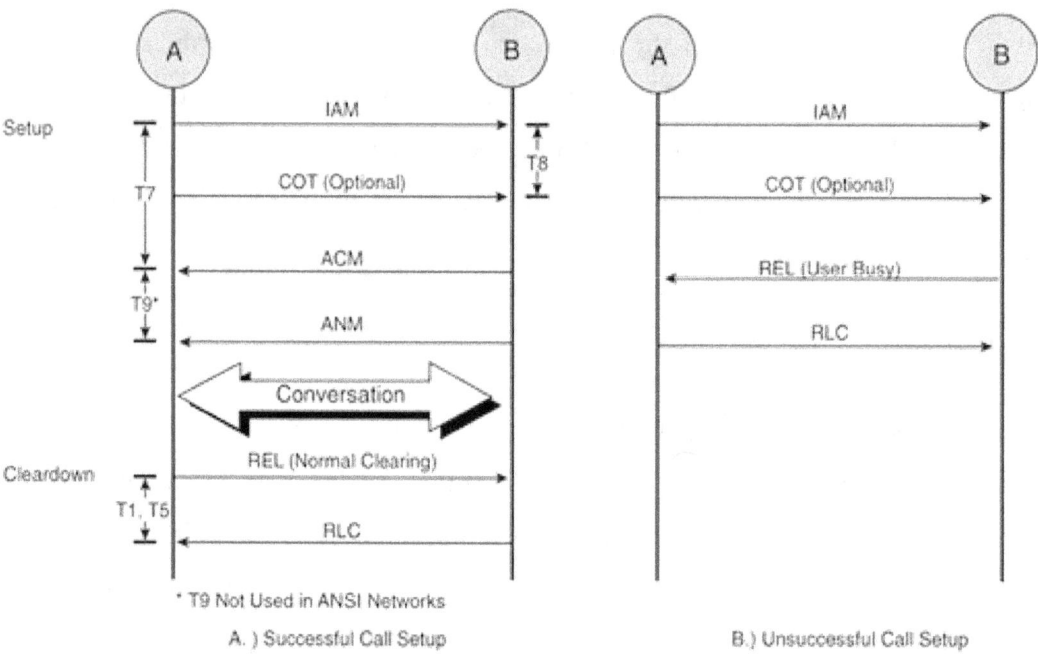

Figure 7-2: Simple ISUP Message Flow

Call Setup

A simple basic telephone service call can be established and released using only five ISUP messages. In Figure 7-2, part A shows a call between SSP A and SSP B. The Initial Address Message (IAM) is the first message sent, which indicates an attempt to set up a call for a particular circuit. The IAM contains information that is necessary to establish the call connection—such as the call type, called party number, and information about the bearer circuit.

Call Release

In Figure 7-2, the call originator at SSP A goes on-hook to end the call. SSP A sends a Release message (REL) to SSP B. The REL message signals the far end to release the bearer channel. SSP B responds with a Release Complete message (RLC) to acknowledge the REL message. The RLC indicates that the circuit has been released.

Unsuccessful Call Attempt

In Figure 7-2, part B shows an unsuccessful call attempt between SSP A and SSP B. After receiving the IAM, SSP B checks the status of the destination line and discovers that it is busy. Instead of an ACM, a REL message with a cause value of User Busy is sent to SSP A, indicating that the call cannot be set up. While this example shows a User Busy condition, there are many reasons that a call set-up attempt might be unsuccessful. For example, call screening at the terminating exchange might reject the call and therefore prevent it from being set up. Such a rejection would result in a REL with a cause code of Call Rejected.

7.5 ISUP Services

ISUP provides bearer services, user terminal services, and supplementary services.

7.5.1 Bearer Services

Bearer service is a low-layer message transfer capability provided by a network. It only indicates ISDN communication capability, and is irrelevant to the type of terminals. Therefore, different terminals can use the same bearer capability.

ISUP supports the following bearer services:
- 64 Kbit/s circuit switching unrestricted
- Speech
- 3.1 kHz audio
- 2 x 64 Kbit/s

- 384 Kbit/s unrestricted
- 1,920 Kbit/s unrestricted

7.5.2 User Terminal Services

User terminal service is application oriented. It includes the communication capability provided by a network and that of a terminal.

Low-layer property indicates the necessary bearer capability of a network, and the property value may be identical with that of the bearer service.

High-layer property indicates the fixed capability of a terminal such as G4 facsimile machine and telephone.

ISUP supports the following user terminal services:
- Telephone
- Intelligent user telegraph
- G2 and G3 facsimile
- G4 facsimile
- Hybrid mode
- Videotext

7.5.3 Supplementary Services

Supplementary services are the extra functions provided by a network to complement bearer services and user terminal services.

Supplementary services must be provided together with a bearer service or user terminal service.

ISUP provides the following supplementary services:
- Call forwarding–unconditional (CFU)
- Call forwarding–busy (CFB)
- Call forwarding–no answer (CFNA)
- Call forwarding–default (CFD)
- Calling number identification presentation (CNIP)
- Calling number identification restriction (CNIR)
- Calling number identification restriction over (CNIR-Over)

- Call waiting (CW)
- Call transfer (CT)

7.5.3.1 Call forwarding–unconditional (CFU)

Call Forwarding—Unconditional (CFU) permits a called subscriber to send incoming calls addressed to the called subscriber's Directory Number to another Directory Number (forward-to number) or to the called subscriber's designated voice mailbox. If this feature is active, calls are forwarded regardless of the condition of the termination. CFU does not impact a subscriber's ability to originate calls. A subscriber may select a forward-to number termination address when it registers the CFU.

7.5.3.2 Call Forwarding-Busy (CFB)

Call Forwarding Busy (CFB) is a service that enables your incoming calls that encounter a busy condition or are not answered, after a customer-designated interval, to be automatically forwarded to another local or long distance phone number where network facilities permit.

7.5.3.3 Call Forwarding–No Answer (CFNA)

Call Forwarding–No Answer (CFNA) permits a called subscriber to have the system send incoming calls addressed to the called subscriber's public identity to another public identity (forward-to identity) or to the called subscriber's designated voice mailbox, when the subscriber fails to answer, or is otherwise inaccessible (including TBD, etc.). CFNA does not apply when the subscriber is considered to be busy.

7.5.3.4 Call Forwarding–Default (CFD)

Call Forwarding–Default (CFD) redirects incoming calls addressed to the called subscriber's Directory Number to the subscriber's designated voice mailbox or to another. Directory Number (forward-to number), when the subscriber is engaged in a call, does not respond to paging, does not answer the call within a specified period after being alerted or is otherwise inaccessible.

7.5.3.5 Calling Number Identification Presentation (CNIP)

Calling Number Identification Presentation (CNIP) provides the number identification of the calling party to the called subscriber. One or two numbers may be presented to identify the calling party.

7.5.3.6 Calling Number Identification Restriction (CNIR)

Calling Number Identification Restriction (CNIR) restricts presentation of that subscriber's Calling Number Identification (CNI) to the called party.

7.5.3.7 Calling number identification restriction over (CNIR-Over)

Calling number identification restriction over (CNIR-Over) is a supplementary service that denies the presentation of CNI to the called party.

7.6 ISUP Messages

This section describes the ISUP message format and types of messages.

7.6.1 ISUP Message Format

The User Data portion of the MTP3 Signaling Information Field contains the ISUP message, identified by a Service Indicator of 5 in the MTP3 SIO field. Each ISUP message follows a standard format that includes the following information:

- CIC— the Circuit Identification Code for the circuit to which the message is related.
- Message Type— the ISUP Message Type for the message (for example, an IAM, ACM, and so on).
- Mandatory Fixed Part— required message parameters that are of fixed length.
- Mandatory Variable Part— required message parameters that are of variable length. Each variable parameter has the following form:
  - Length of Parameter
  - Parameter Contents
- Optional Part— Optional fields that can be included in the message, but are not mandatory. Each optional parameter has the following form:
  - Parameter Name
  - Length of Parameter
  - Parameter Contents

The message structure provides a great deal of flexibility for constructing new messages. Each message type defines the mandatory parameters that are necessary for constructing a message. The mandatory fixed variables do not contain length information because the ISUP standards specify them to be a fixed length. Because the mandatory variable parameters are of variable lengths, pointers immediately follow the mandatory fixed part to point to the beginning of each variable parameter. The pointer value is simply the number of octets from the pointer field to the variable parameter length field.

In addition to the mandatory fields, each message can include optional fields. The last of the pointer fields is a pointer to the optional part. Optional fields allow information to be included or omitted as needed on a per-message basis. The optional fields differ based on variables such as the call type or the supplementary services involved. For example, the Calling Party Number (CgPN) field is an optional parameter of the IAM, but is usually included to provide such services as Caller ID and Call Screening.

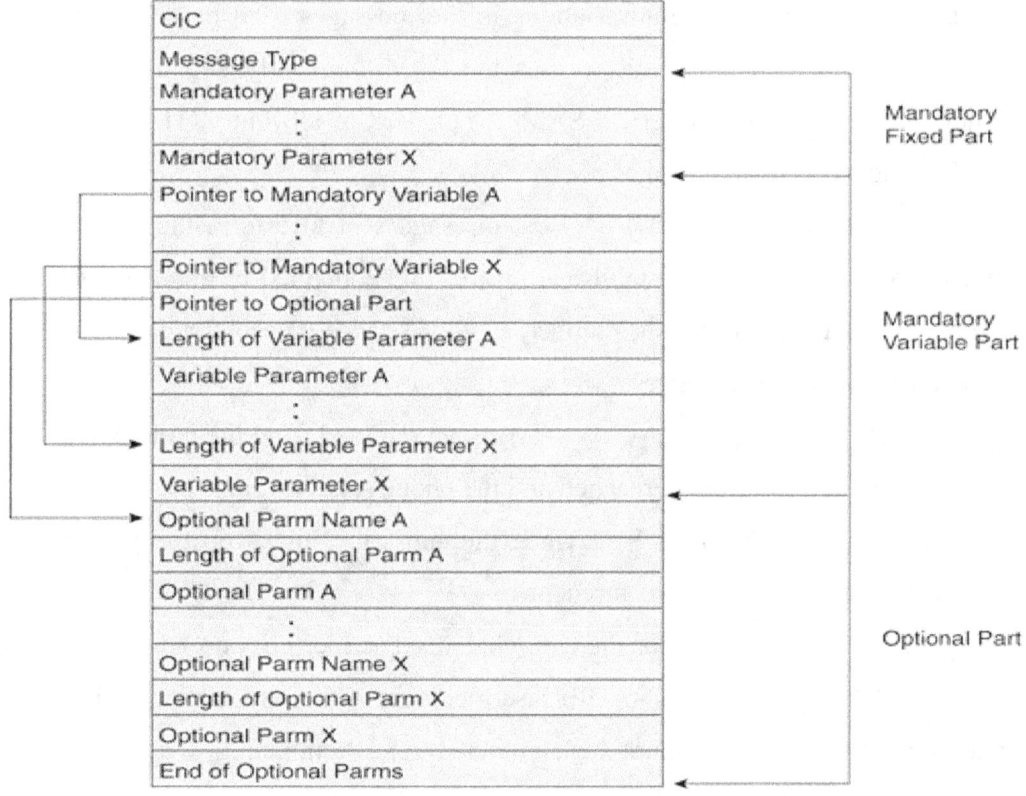

Figure 7-3: ISUP Message Format

A single message can include many optional parameters. The optional part pointer field only points to the first parameter. Because the message might or might not include the parameters, and because the parameters can appear in any order, the first octet includes the name of each parameter in order to identify it. The parameter length follows the name to indicate how many octets the parameter contents include. When the parameter name is coded as zero, it signals the end of the optional parameters. During parsing of an incoming ISUP message, optional parameters are processed until the end of optional parameters marker is reached. If the message does not have any optional parameters, the pointer to the optional part is coded to zero.

7.6.2 Message types

An ISUP message contains a fixed header containing the circuit identification code and the ISUP message type, followed by a fixed-length part and optional variable-length part that are dependent on the type of message being sent. ISUP messages can be sent using the services of the Message Transfer Part, or, less often, the Signaling Connection Control Part. These messages are transmitted in various stages of call setup and teardown. There are slight variations in the messages used based on the individual network. For example, Europe uses the SAM frequently and the COT message more rarely. In North America, SAM is not used at all, but COT is used more often. The most common messages are:

- Initial Address Message (IAM) — First message sent to inform the partner switch that a call has to be established on the CIC contained in the message. Contains the called and calling number, type of service (speech or data) and many more optional parameters.
- Address Complete Message (ACM) — Message returned from the terminating switch when the subscriber is reached and the phone starts ringing.
- Answer Message (ANM) — Sent when the subscriber picks up the phone. Normally charging starts at this moment.
- Release (REL) — Sent to clear the call when a subscriber goes on hook. This is also sent (without a preceding Release message) if the terminating switch determines that the call cannot be completed. The terminating switch also sends a Cause Value to explain the reason for the failure, *e.g.*, "User busy".

- <u>Release complete</u> (RLC) — Acknowledgement of the release – the timeslot is idle afterwards and can be used again. [20]

7.6.2.1 Initial Address Message (IAM)

The IAM contains the information needed to set up a call. For a basic call, it is the first message sent and is typically the largest message in terms of size. Figure 7-4 shows the mandatory fields that the message includes. In addition to the mandatory fields, the ITU-T Q.764 lists more than 50 optional parameters that can be included in the IAM. The mandatory parameters for ITU and ANSI are the same, with the exception of the Transmission Medium Requirements parameter. In ANSI networks, the User Service Info field is used instead.

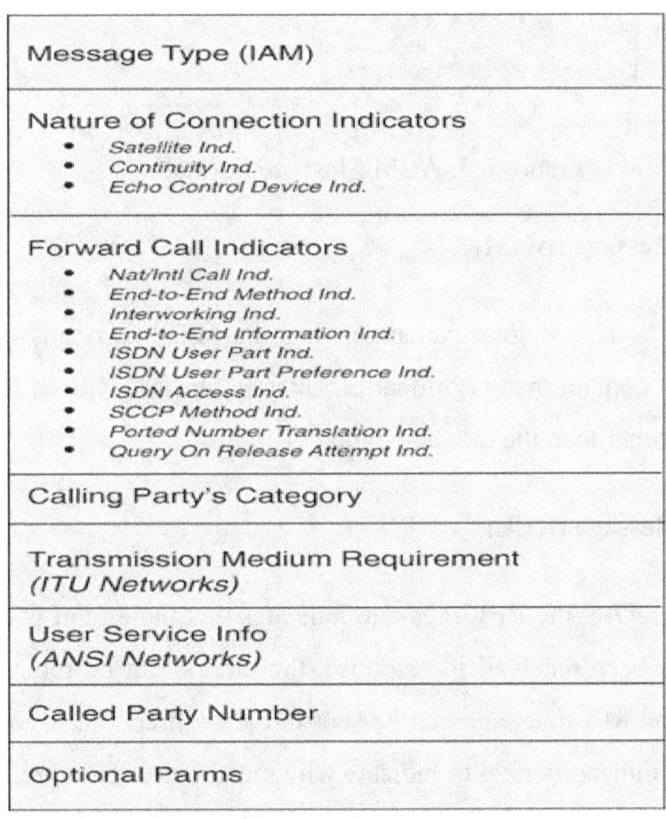

Figure 7-4: IAM Message Format

7.6.2.2 Address Complete Message (ACM)

As shown in Figure 7-5, a destination node sends the ACM to indicate that a complete CdPN has been received. When enbloc signaling is used to set up the call, the ACM is sent after receiving the IAM; when overlap signaling is used, it is sent after the last SAM is received. In addition to indicating the successful reception of the CdPN, the ACM sends Backward Call Indicators (BCI) to signal information about the call setup. It is not mandatory for an ACM to be sent when setting up a call. It is permissible to send an ANM after receiving an IAM; this is sometimes referred to as "fast answer."

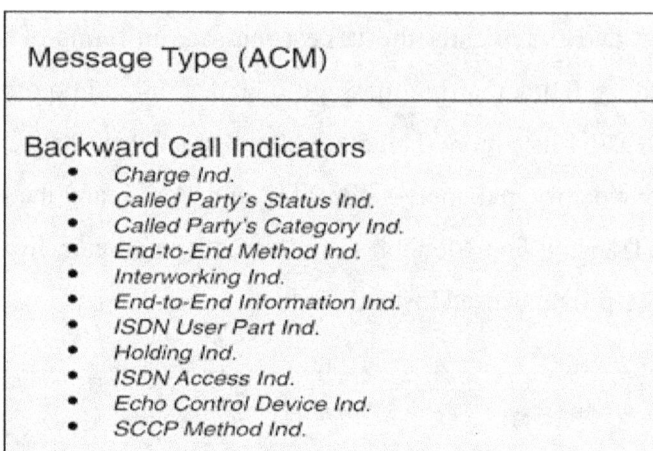

Figure 7-5: ACM Message Format

7.6.2.3 Answer Message (ANM)

The ANM is sent to the previous exchange when the called party answers (off-hook). Although it might contain many optional parameters, the ANM does not contain any mandatory fields other than the message type.

7.6.2.4 Release Message (REL)

As shown in Figure 7-6, the REL message indicates that the circuit is being released. When a RLC has been received in response, the circuit can be returned to the idle state for reuse. The REL message can be sent in either direction. It contains a single mandatory Cause Indicators field to indicate why the circuit is being released.

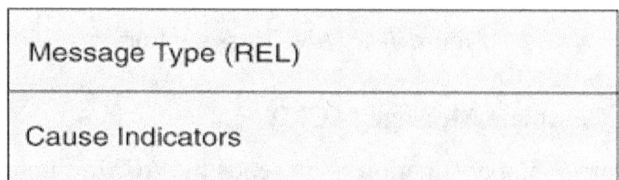

Figure 7-6: REL Message Format

7.6.2.5 Release Complete Message (RLC)

The RLC message is sent to acknowledge a REL message. Upon receipt of an RLC, a circuit can return to the idle state. [2]

7.6.3 Subscriber Interface Message Tracing [16]

A. Call Between Same Operators:

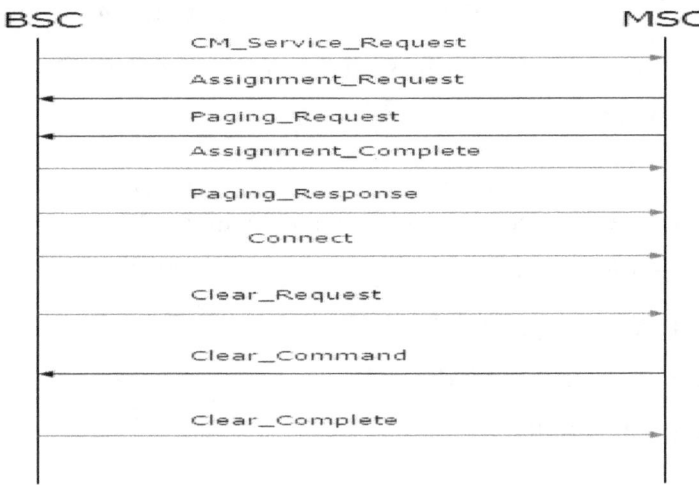

Figure 7-7: Interface Message Tracing in Same Operator

B. Call Between Different Operators:

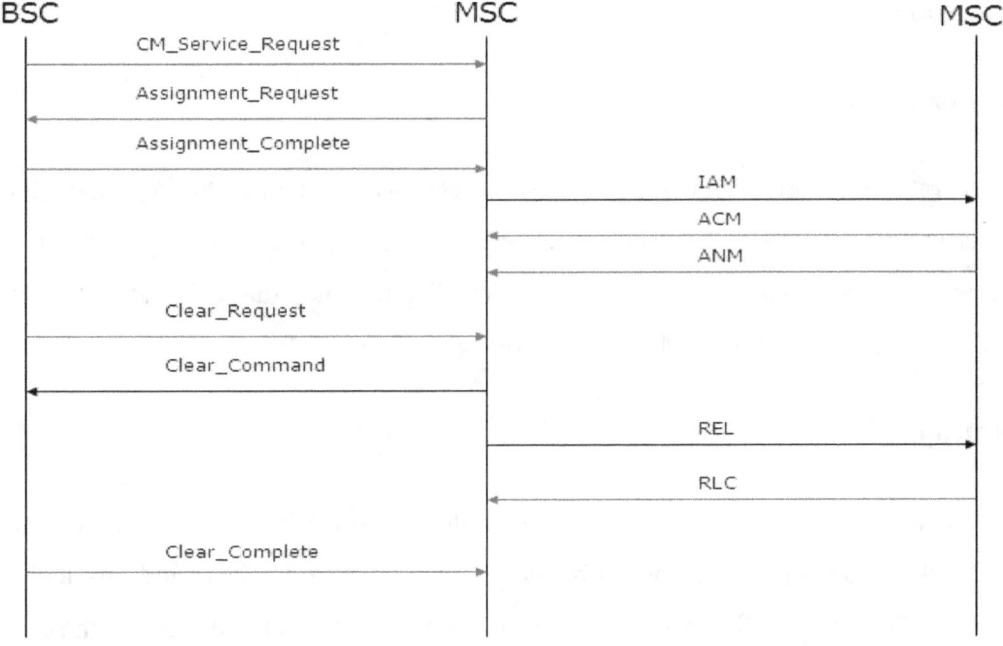

Figure 7-8: Interface Message Tracing in Different Operator

7.7 ISUP and Local Number Portability

Local number portability (LNP) for fixed lines, and full mobile number portability (FMNP) for mobile phone lines, refers to the ability to transfer either an existing fixed-line or mobile telephone number assigned by a local exchange carrier (LEC) and reassign it to another carrier. In most cases, there are limitations to transferability with regards to geography, service area coverage and technology. [21]

The actual use of LNP in the network exists today, but only to a small degree. It is being expanded in phases and will take some time before it is ubiquitous across all networks and locations.

Currently, four mechanisms are defined for implementing NP:
- All Call Query (ACQ)
- Query on Release (QOR)
- Dropback or Release to Pivot (RTP)
- Onward Routing (OR)

All Call Query (ACQ)

The operator that originates the call always checks a centralized database and obtains the route to the call.

Query on Release (QoR)

The operator that originates the call first checks with the operator to which the number initially belonged, the donor operator. The donor operator verifies the call and informs that it no longer possesses the number. The operator that originates the call then checks the centralized database, as is done with ACQ.

Call Dropback

Also known as Return to Pivot (RoP). The operator that originates the call first checks with the donor operator. The donor operator checks its own database and provides a new route. The operator that originates the call then uses this route to forward the call. No central database is consulted.

Onward Routing (OR)

The operator that originates the call checks with the donor operator. The donor operator checks its own database and obtains a new route. The operator to which the number was designated routes itself the call to the new operator. This model is called indirect routing. [21]

7.8 Interworking with ISDN

ISDN uses a common channel (the D channel) for access signaling; this compliments the common channel network signaling ISUP uses and provides a complete digital signaling path between end users when ISDN is used for network access and ISUP is used throughout the core network.

A correlation exists between the ISDN messages from the user premises and the ISUP messages on the network side of the call. Figure 7-9 illustrates this correlation using an ISDN-to-ISDN call over an ISUP facility.

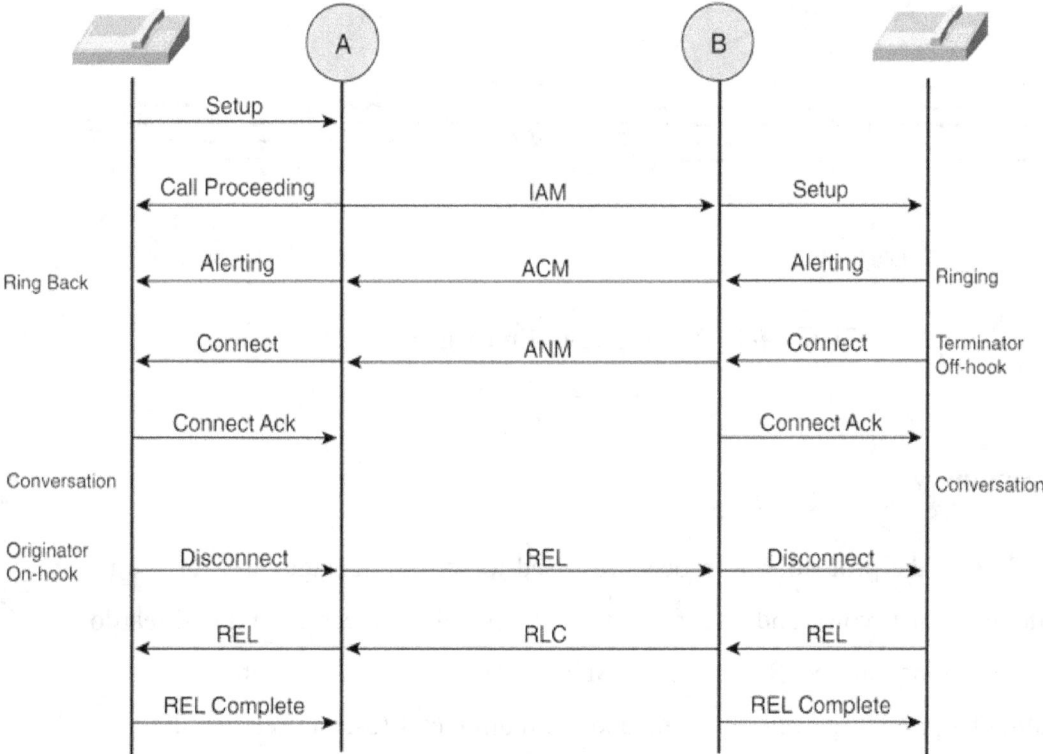

Figure 7-9: ISUP-ISDN Interworking

7.9 End-to-End Signaling

The ability to perform end-to-end signaling is accomplished using ISDN access signaling and ISUP network signaling. End-to-end signaling is the passing of information across the network that is only pertinent to the two communicating endpoints. Generally, this means that the two phone users are connected across the network. The network itself can be viewed as a communications pipe for the user information.

There are two different methods for end-to-end signaling over ISUP: the Pass Along Method (PAM) and the SCCP Method. As shown in Figure 7-10, PAM exchanges end-to-end signaling by passing along information from one node to the next, based on the physical connection segments.

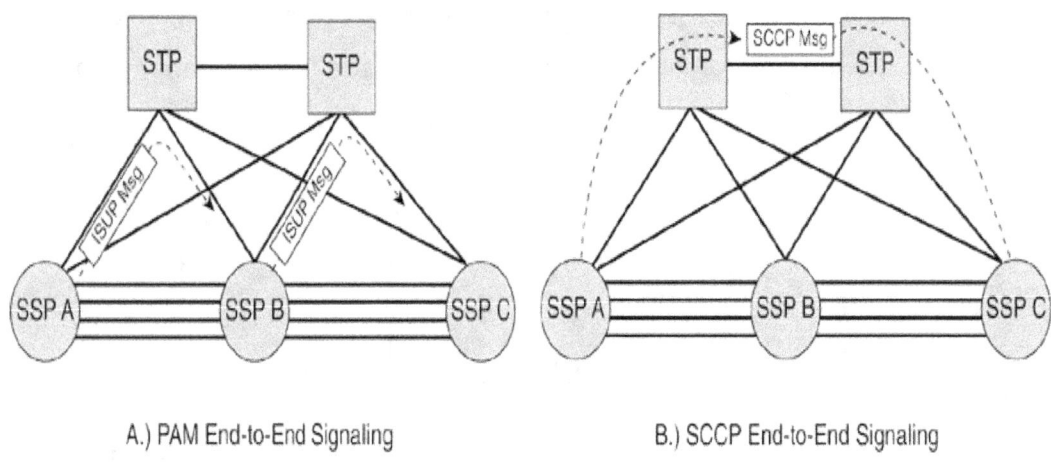

A.) PAM End-to-End Signaling   B.) SCCP End-to-End Signaling

Figure 7-10: ISUP End-to-End Signaling

7.10 Summary

ISUP defines the protocol and procedures used to set-up, manage, and release trunk circuits that carry voice and data calls. Call setup, call connection and call teardown helps describing the ISUP call flow. ISUP helps interface well with ISDN access signaling by providing event mapping and facilitating end-to-end user signaling.

ISUP provides bearer services, user terminal services, and supplementary services. ISUP messages are transmitted in various stages of call setup and teardown. Initial

Address Message (IAM), Subsequent Address Message (SAM), Address Complete Message (ACM), Answer Message (ANM), Release (REL), Release complete (RLC) are the types of messages in ISUP.

ISUP uses a CIC identifier in each message to correlate the signaling with the correct circuit. ISUP also provides a set of maintenance messages for diagnostics and maintenance of ISUP facilities.

Chapter 8

Transaction Capabilities Application Part (TCAP)

8.1 Introduction

With the development of telecommunication networks, more services are demanded. Such services include the intelligent services like freephone (FPH) and virtual private network (VPN), as well as the operation, administration, maintenance and provision (OAM&P) and mobile application part (MAP). These services and applications are irrelevant to call control. That is, message transfer functions are separated from call control functions. They are provided on the basis of the correlation between. [16]

1. Exchanges
2. Exchanges and network service centers
3. Subscribers and network service centers

To address these demands, the transaction capability (TC) protocol is applied. Transaction capabilities are functions that control non-circuit-related information transfer between two or more signaling nodes through the SS7. They serve as the interface between several applications and one particular service. The TC protocol provides general standards for the applications as a whole instead of for a particular application. The TC consists of transaction capability application part (TCAP) and intermediate service part (ISP). The former corresponds to Layer 7 of the OSI model, and the latter, Layers 4–6. In the CDMA system only TCAP is involved. That is, TCAP is directly involved in data transfer.

Figure 8-1 shows the position of TCAP in the SS7 network.

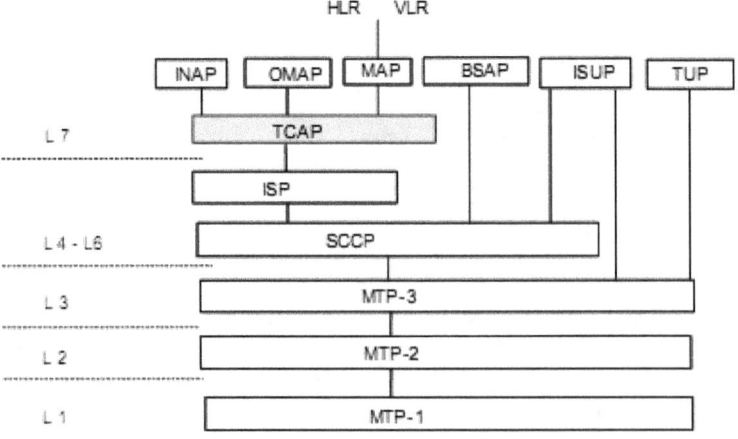

Figure 8-1: Position of TCAP in the SS7 network

8.2 TCAP Structure

TCAP is divided into

1. Component sublayer (CSL): responsible for operation administration

2. Transaction sublayer (TSL): responsible for transaction administration

The CSL communicates with TC user over TC primitive interface and with the TSL over TR primitive interface. [16]

Figure 8-2 shows the structure of TCAP.

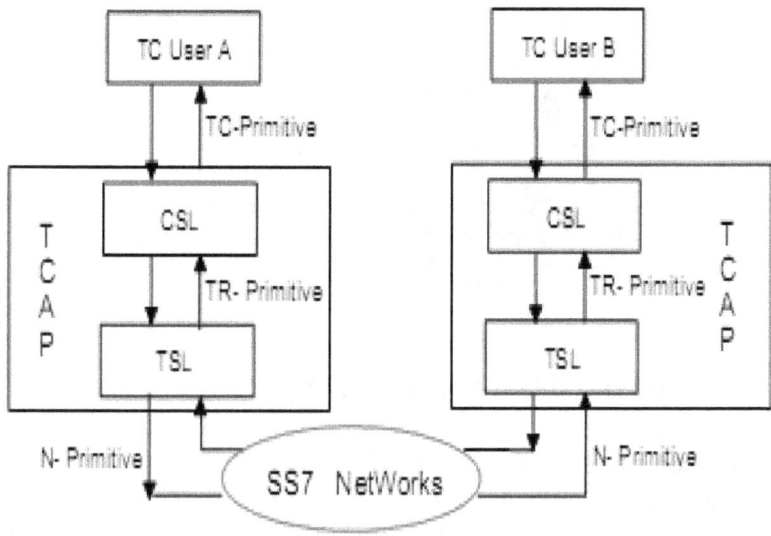

Figure 8-2: The structure of TCAP.

8.3 Role of TCAP in Call Control

TCAP is used to provide information to SSPs. This information is often used to enable successful call completion, but TCAP is not involved in the actual call-setup procedures. The protocol's circuit-related portion, such as ISUP and TUP, perform the call setup. This interaction between the service information provided by TCAP and the circuit-related protocol that performs the call setup occurs at the application level, not at the SS7 protocol layer. [2]

8.4 TCAP within the SS7 Protocol Stack

As shown in figure 8-3 TCAP is at level 4 of the SS7 protocol stack. It depends upon the SCCP's transport services because TCAP itself does not contain any transport information. First, SCCP must establish communication between services before TCAP data can be delivered to the application layer.

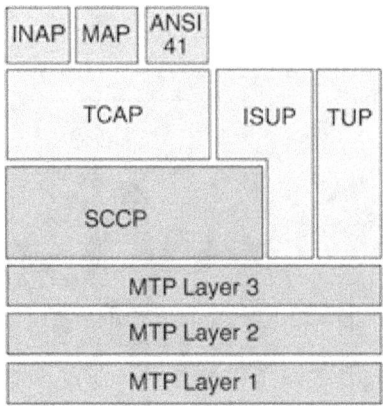

Figure 8-3: TCAP within the SS7 Stack

8.5 Transaction and Component Sublayers

The TCAP message is composed of two main sections: the transaction sublayer and the component sublayer. A transaction is a set of related TCAP messages that are exchanged between network nodes. The transaction portion identifies the messages that belong to the same transaction using a Transaction Identifier (TRID). The message's component portion contains the actual instructions, or "operations," that are being sent to the remote application.

8.6 Message Layout

Now that we have examined in detail how each of the TCAP data elements are constructed, let's take a look at how they are assimilated into a message. There are three distinct sections into which a message is divided: the transaction, dialogue, and component portions. The dialogue portion of the message is optional. Figure 8-4 shows the complete structure of a TCAP message within the context of its supporting SS7 levels. [2]

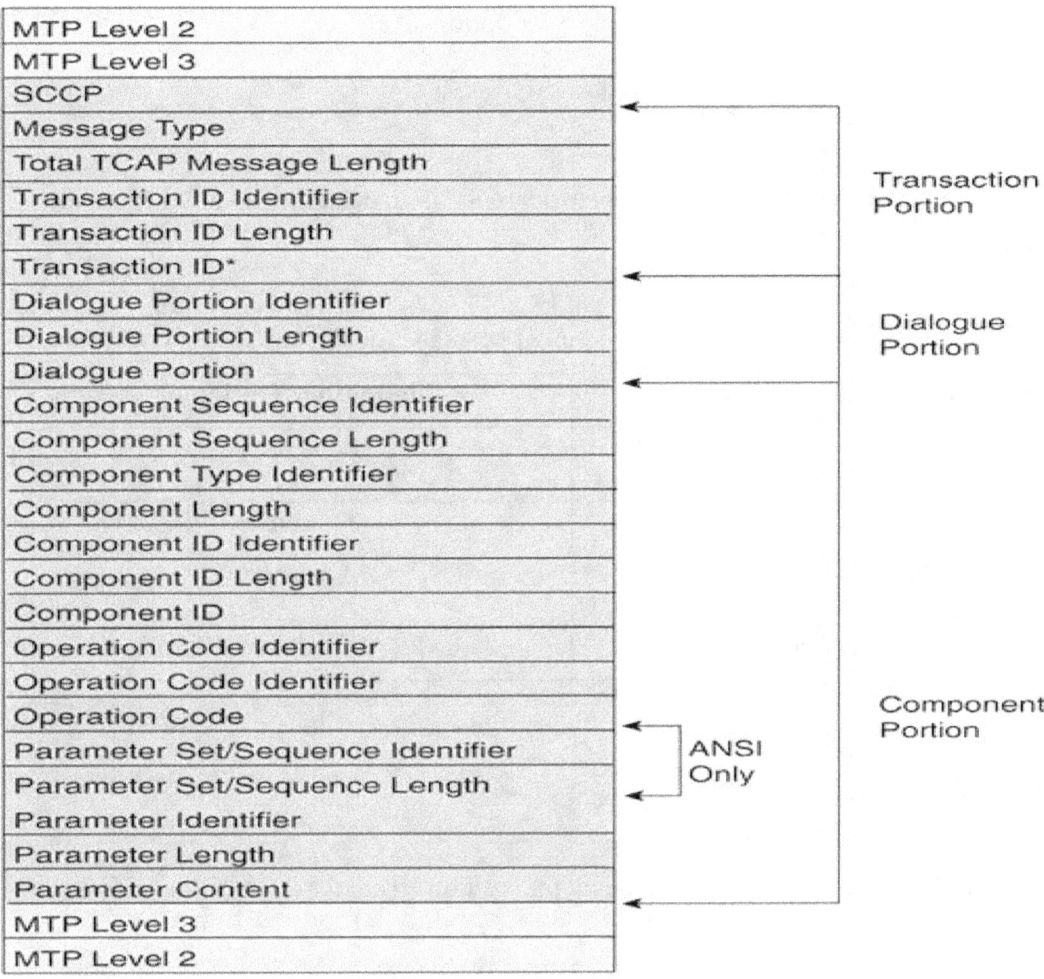

*0, 1, or 2 Transaction ID fields may be included, depending on message type.

Figure 8-4: TCAP Message Structure

8.7 ITU Protocol Message Contents

The definition of each message type indicates a set of fields that comprise the message. While some fields are mandatory, others are optional. As specified by Q.773, the standard set of ITU messages includes:

- Unidirectional
- Begin
- End
- Continue
- Abort

The following sections describe these messages, the fields that are included in each one, and indicate which fields are mandatory or optional.[2]

8.7.1 Unidirectional Message

The Unidirectional Message is sent when no reply is expected. Table 8-1 lists the message contents.

Table 8-1: Unidirectional Message Fields

Unidirectional Message Fields	Mandatory/Optional
Message Type Total Message Length	Mandatory
Dialogue Portion	Optional
Component Portion Tag Component Portion Length	Mandatory
One or More Components	Mandatory

8.7.2 Begin Message

The Begin Message is sent to initiate a transaction. Table 8-2 lists the message contents.

Table 8-2: Begin Message Fields

Begin Message Fields	Mandatory/Optional
Message Type Total Message Length	Mandatory
Originating Transaction ID Tag Transaction ID Length Transaction ID	Mandatory
Dialogue Portion	Optional
Component Portion Tag Component Portion Length	Optional
One or More Components	Optional

8.7.3 End Message

The End Message is sent to end a transaction. Table 8-3 lists the message contents.[2]

Table 8-3: End Message Fields

End Message Fields	Mandatory/Optional
Message Type Total Message Length	Mandatory
Destination Transaction ID Tag Transaction ID Length Transaction ID	Mandatory
Dialogue Portion	Optional
Component Portion Tag Component Portion Length	Optional
One or More Components	Optional

8.7.4 Continue Message

The Continue Message is sent when a transaction has previously been established and additional information needs to be sent without ending the transaction. Table 8-4 lists the message contents.

Table 8-4: Continue Message Fields

Continue Message Fields	Mandatory/Optional
Message Type Total Message Length	Mandatory
Originating Transaction ID Tag Transaction ID Length Transaction ID	Mandatory
Destination Transaction ID Tag Transaction ID Length Transaction ID	Mandatory
Dialogue Portion	Optional
Component Portion Tag Component Portion Length	Optional
One or More Components	Optional

8.7.5 Abort Message

The Abort Message is sent to terminate a previously established transaction. Table 8-5 lists the message contents.

Table 8-5: Abort Message Fields

Abort Message Fields	Mandatory/Optional
Message Type Total Message Length	Mandatory
Destination Transaction ID Tag Transaction ID Length Transaction ID	Mandatory
P-Abort Cause Tag P-Abort Cause Length P-Abort Cause	Optional
Dialogue Portion	Optional

TCAP provides a standard mechanism for telephony services to exchange information across the network. It is designed to be generic so it can interface with a variety of services.

TCAP resides at Level 4 of the SS7 protocol and depends on SCCP's transport services. It is comprised of a transaction sublayer and a component sublayer. The transaction sublayer correlates the exchange of associated messages, while the component sublayer handles the remote operation requests.

All information elements in the TCAP message are defined and encoded using the syntax and BER of ASN.1. The ITU Q.771—Q.775 series of specifications defines the TCAP protocol. Specifications such as the ETSI.300.374 INAP series build on the ITU Q Series Recommendations to provide additional information needed for implementing network services.[2]

TCAP traffic on telephony signaling networks has increased in recent years because of an increase in services such as LNP, Calling Name Delivery, and Short Messaging Service (SMS), which rely on TCAP communication. This upward trend is likely to continue as IN services are more widely deployed, thereby making TCAP an increasingly important component in the role of network services.

8.8 Summary

TCAP provides a standard mechanism for telephony services to exchange information across the network. It is designed to be generic so it can interface with a variety of services.

TCAP resides at Level 4 of the SS7 protocol and depends on SCCP's transport services. It is comprised of a transaction sublayer and a component sublayer. The transaction sublayer correlates the exchange of associated messages, while the component sublayer handles the remote operation requests.

TCAP traffic on telephony signaling networks has increased in recent years because of an increase in services such as LNP, Calling Name Delivery, and Short Messaging Service (SMS), which rely on TCAP communication. This upward trend is likely to continue as IN services are more widely deployed, thereby making TCAP an increasingly important component in the role of network services

Chapter 9

Mobile Application Part (MAP)

9.1 Introduction

The Mobile Application Part (MAP) is an SS7 protocol which provides an application layer for the various nodes in GSM and UMTS mobile core networks and GPRS core networks to communicate with each other in order to provide services to mobile phone users. The Mobile Application Part is the application-layer protocol used to connect the distributed switching elements, called mobile switching centers (MSCs) with a master database called the Home Location Register (HLR). The HLR dynamically stores the current location and profile of a mobile network subscriber. The HLR is consulted during the processing of an incoming call. Conversely, the HLR is updated as the subscriber moves about the network and is thus serviced by different switches within the network. [15]

Mobile application part (MAP) is used to allow the GSM network nodes within the Network Switching Subsystem (NSS) to communicate with each other to provide services, such as roaming capability, text messaging (SMS), and subscriber authentication. MAP provides an application layer on which to build the services that support a GSM network. This application layer provides a standardized set of operations. MAP is transported and encapsulated with the SS7 protocols MTP, SCCP, and TCAP.

MAP has been evolving as wireless networks grow, from supporting strictly voice, to supporting packet data services as well. The fact that MAP is used to connect NexGen elements such as the Gateway GPRS Support node (GGSN) and Serving Gateway Support Node (SGSN) is a testament to the sound design of the GSM signaling system.

Figure 9-1 presents the interfaces between these entities.

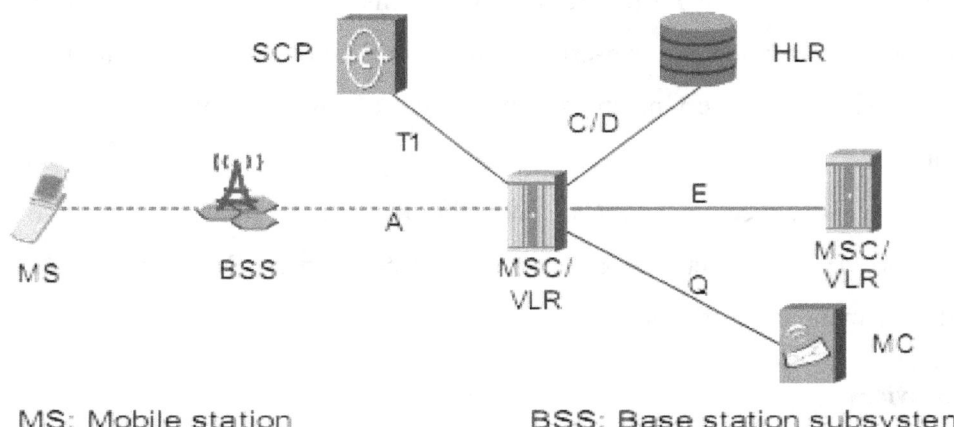

Figure 9-1: Mobile network architecture

Except for the A interface, all interfaces in the Mobile network can transmit MAP messages.

The following describes these interfaces.

- A interface

    The A interface is between the network subsystem and base station subsystem. This interface carries messages related to MS management, BTS management, mobility management, call processing, and so on.

- B interface

    B interface is VLR to MSC interface. Through this interface, the MSC requests location information from the VLR and notifies the VLR to update the location information of a MS. This interface also carries supplementary services operation messages.

- C interface

    C interface is MSC to HLR interface. In a mobile terminated call, the GMSC obtains roaming number from the HLR through the C interface. In a mobile terminated short message service, the MC obtains, over this interface, the number of serving MSC from HLR through GMSC.

- D interface

    D interface is VLR to HLR interface. Over this interface, VLR and HLR exchange MS location and subscriber management information to ensure that the subscribers in the serving area can make and receive calls normally.

- E interface

    E interface is between two MSCs, controlling the handoff of MSs between to neighbor MSCs. E interface carries messages between the MSCs to initiate and implement handoff operations. [16]

9.2 MAP Services

The primary facilities provided by MAP are:

- Mobility Services: location management (roaming), authentication, managing service subscription information, fault recovery,
- Operation and Maintenance: subscriber tracing, retrieving a subscriber's IMSI
- Call Handling: routing, managing calls whilst roaming, checking that a subscriber is available to receive calls
- Supplementary Services
- Short Message Service
- Packet Data Protocol (PDP) services for GPRS: providing routing information for GPRS connections
- Location Service Management Services: obtaining the location of subscribers [15]

9.2.1 Mobility Services

Mobility management operations can be divided into the following categories:

- Location Management
- Paging and Search
- Access Management
- Handover
- Authentication Management
- Security Management
- IMEI Management

- Subscriber Management
- Identity Management
- Fault Recovery

9.2.2 Operation and Maintenance

Operation and maintenance can be divided into the following categories:

- Subscriber Tracing
- Miscellaneous

9.2.3 Call Handling

The call handling procedures primarily retrieve routing information to allow mobile terminating calls to succeed. When a mobile originating or a mobile terminating call has reached the destination MSC, no further MAP procedures are required.

Other procedures performed by MAP's call handling routines include the restoration of call control to the Gateway Mobile Switching Center (GMSC) if the call is to be forwarded. In addition, the call handling routing processes the notification that the remote user is free for the supplementary service message call completion to busy subscribers (CCBS).

Call handling does not have subcategories of operations; it simply has the following two operations:

- sendRoutingInfo
- provideRoamingNumber

9.2.4 Supplementary Services

Supplementary service includes the following operations:

- registerSS
- eraseSS
- activateSS
- deactivateSS
- interrogateSS

- registerPassword
- getPassword

In addition to these supplementary services, the following operations are considered unstructured supplementary services:

- processUnstructuredSS-Request
- unstructuredSS-Request
- unstructuredSS-Notify

9.2.5 Short Message Service (SMS)

SMS provides paging functionality for alphanumeric messages of up to 160 characters to be exchanged with other GSM users. The network itself can also generate messages and broadcast to multiple MSs or to a specific MS. For example, a welcome message can be sent to a subscriber when he or she roams onto a new network; in addition, it can provide useful information, such as how to retrieve voicemail. The SMS service also transfers ring tones and logos to the MS.

The SMS slightly blurs the image of the user traffic being separate from signaling because, in a sense, the messages are user traffic; they are for human processing (written and read), rather than for communication between network entities.

The SMS does not have subcategories. It has the following operations:

- forwardSM
- sendRoutingInfoForSM
- reportSMDeliveryStatus
- readyForSM
- alertServiceCentre
- informServiceCentre

9.3 MAP Operations

Implementation of each MAP function contains several operations. Each operation is defined by a set of elements including:

- Operation name, code, and type
- Invoke parameter
- Success parameter
- Failure code and parameter
- Linked operations allowed

MAP operations are classified into four categories:

- Category 1 operations: Report is required regardless of the operation result. In the case of a successful operation, the result is reported; in the case of an unsuccessful operation, the error is reported.
- Category 2 operations: Report is required only in the case of operation failure.
- Category 3 operations: Report is required only in the case of operation success.
- Category 4 operations: Report is not required.

For the sake of security, when MAP originates a remote operation, the operation time limit must be specified. If no report is received in the time limit, processing is as follows:

- For categories 1 and 3, it is considered operation failure
- For categories 2 and 4, it is considered success.

Currently, category 4 only contains OANSWER and TANSWER operations. Other operations are all classified to category 1.

9.4 MAP Messages

9.4.1 Format of MAP Messages

In the SS7, MAP messages are transmitted as part of TCAP messages.

Figure 9-2 shows the structural relation between MAP and MTP messages.

Figure 9-2: Structural relation between MAP and MTP messages

MAP messages are coded in ASN.1 format. The message type is in one to one correspondence with the operation code in the TCAP component.

In message transmission, one MAP message corresponds to one invoke ID. The invoke ID is the unique identifier of a MAP message. Thus, a TCAP component can be translated to a MAP message based on the invoke ID. [16]

9.5 Mobility Management and Call Processing

This section provides an introductory overview of mobility management (i.e., allowing a subscriber to roam) and call processing (the setting up and clearing down of calls) in GSM networks.

Mobility management entails keeping track of the MS while it is on the move. The mobility management procedures vary across three distinct scenarios, namely:

- MS is turned off
- MS is turned on but is idle
- MS has an active call

In the first scenario, when it cannot be reached by the network because it does not respond to the paging message, the MS is considered to be in the turned-off state. In this scenario, the MS obviously fails to provide any updates in relation to changes in Location Area (LA), if any exist. In this state, the MS is considered detached from the system (IMSI detached).

In the second scenario, the MS is in the ready state to make or receive calls. The system considers it attached (IMSI attached), and it can be successfully paged. While on the move, the MS must inform the system about any changes in LA; this is known as location updating.

In the third scenario, the system has active radio channels that are allowed to the MS for conversation/data flow. The MS is required to change to new radio channels if the quality of current channels drops below a certain level; this is known as handover. The MSC (sometimes BSC) makes the decision to handover an analysis of information that is obtained real-time from the MS and BTS.

All operations revolve around the three scenarios presented above. The rest of this chapter examines these operations in more detail, beginning with simple operations: paging, IMSI detach/attach. Following, more complex operations are presented, such as location update, call handover, mobile terminated call, mobile originated call, and mobile-to-mobile call.

Location Update

Location updating is the mechanism that is used to determine the location of an MS in the idle state. The MS initiates location updating, which can occur when:
- The MS is first switched on
- The MS moves within the same VLR area, but to a new LA
- The MS moves to a new VLR area
- A location updated timer expires

Mobile Terminated Call (MTC)

In the case of an MTC, a subscriber from within the PSTN dials the mobile subscriber's MSISDN. This generates an ISUP IAM message (it also could potentially be TUP as Level 4) that contains the MSISDN as the called party number. The ISDN (i.e., PSTN) routes the call to the GMSC in the PLMN, based on the information contained in the MSISDN (national destination code and the country code).

The GMSC then identifies the subscriber's HLR based upon the MSISDN and invokes the MAP/C operation Send Routing Information (SRI) towards the HLR to locate the MS. The SRI contains the MSISDN. The HLR uses the MSISDN to obtain the IMSI.

Because of past location updates, the HLR already knows the VLR that currently serves the subscriber. The HLR queries the VLR using the MAP/D operation Provide Roaming Number (PRN) to obtain the MSRN. The PRN contains the subscriber's IMSI.

The VLR assigns a temporary number known as the mobile station roaming number (MSRN), which is selected from a pool, and sends the MSRN back in an MAP/D MSRN Acknowledgement to the HLR.

The HLR then passes the MSRN back to the GMSC in a MAP/C Routing Information Acknowledgement message. To the PSTN, the MSRN appears as a dial able number.

Since the GMSC now knows the MSC in which the MS is currently located, it generates an IAM with the MSRN as the called party number. When the MSC receives the IAM, it recognizes the MSRN and knows the IMSI for which the MSRN was allocated. The MSC then returns the MSRN to the pool for future use on another call.

The MSC sends the VLR a MAP/B Send Information message requesting information, including the called MS's capabilities, services subscribed to, and so on. If the called MS is authorized and capable of taking the call, the VLR sends a MAP/B Complete Call message back to the MSC.

The MSC uses the LAI and TMSI received in the Complete Call message to route a BSSMAP Page message to all BSS cells in the LA.

Air interface signaling is outside the scope of this book.

Figure 9-3 shows the sequence of events involved in placing an MTC.

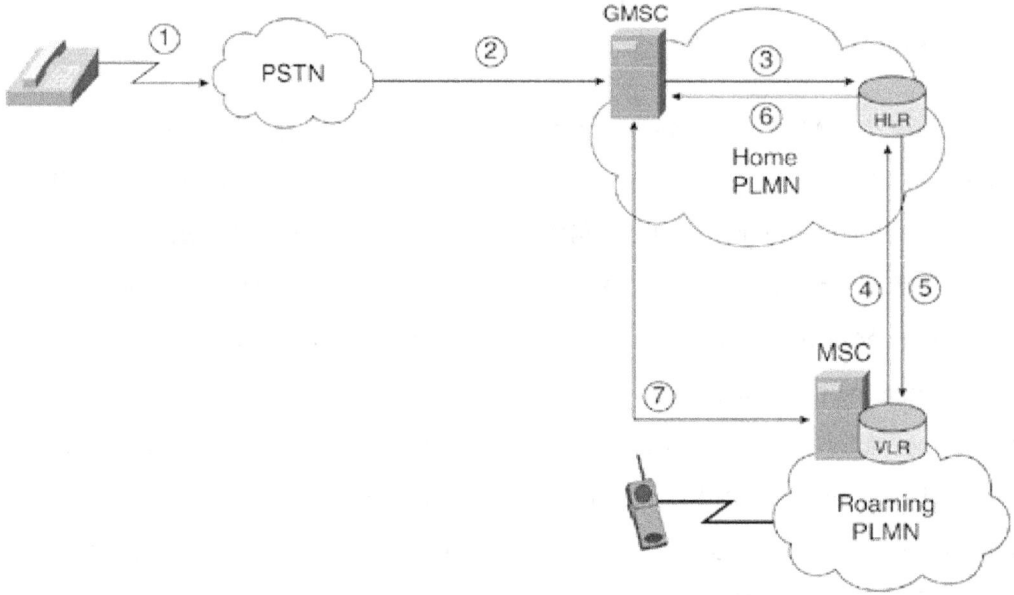

Figure 9-3: Placing an MTC

In Figure 9-3, the sequence of events involved in placing an MTC is as follows:

1. The calling subscriber uses the MSISDN to dial the mobile subscriber.
2. The MSISDN causes the call to be routed to the mobile network gateway MSC (GMSC).

3. The GMSC uses information in the called number digits to locate the mobile subscriber's HLR.
4. The HLR has already been informed about the location (VLR address) for the mobile subscriber; it requests a temporary routing number to allow the call to be routed to the correct MSC.
5. The MSC/VLR responds with a temporary routing number that is only valid for the duration of this call.
6. The routing number is returned to the GMSC.
7. The call is made using ISUP (or TUP) signaling between the GMSC and the visited MSC.

If the calling subscriber were in the same PLMN as the called party (internal MS-to-MS call), steps 2 and 3 would not be required.

9.6 Summary

Modern cellular networks are digital and use SS7 for communication between network entities. GSM is the most popular digital cellular standard. GSM management call control, subscriber mobility, and text messaging (SMS) use a SS7 subsystem known as MAP.

MAP primary use is to allow calls to be delivered to mobile subscribers. MAP keeps track of a mobile subscriber and provides other functionality, including allowing mobile subscribers to send alphanumeric two-way text between handsets; this is known as SMS. MAP also provides mobile operators with the functionality to manage a subscriber's subscription so that services can be added and removed in real-time.

Chapter 10

Practical Implementation of SS7 Signaling Link

10.1 Introduction

We have already described SS7 network architecture and protocols vividly. Now, we will analysis practical SS7 signaling configuration of core network for an operator [22]. Figure 10-1 shows some SPs such as MSC, HLR and IN for a telecom operator.

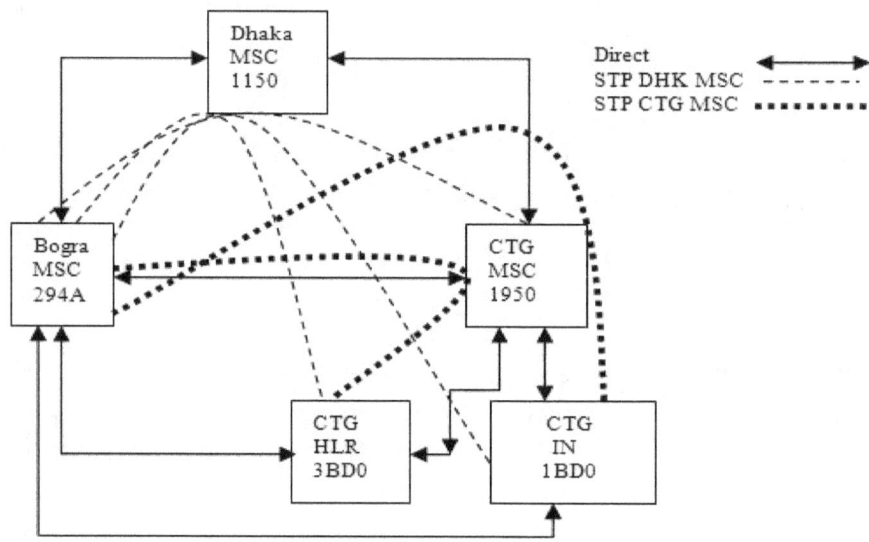

Figure 10-1: Different SPs location showing direct and STP connection of a telecom core network.

10.2 Explanation

Let, there is a telecom operator which has three core network switches (MSC) around the country named as Bogra MSC (BOG_MSC), Dhaka MSC (DHK_MSC) and Chittagong MSC (CTG_MSC). All MSC are known as SP. The Home location register (HLR) and Intelligent network (IN) of the operator are also in CTG region under CTG_MSC. HLR and IN are also treated as different SP. These all Entities are shown in Fig. 3. There is a direct connection between BOG_MSC to all other SPs as well as STP connection (Via Path) except for DHK_MSC as there is no STP link. There is also direct connection between CTG_MSC and CTG_ HLR & CTG_IN.

Let, these arrangements of data are configured in BOG_MSC. We explain here SS7 signaling configuration for BOG_MSC perspective only.

Table 10-1: MTP Point Code Information

DSP Index	DSP Name	OPC	DPC
1	DHK_MSC	294A	1150
2	CTG_MSC	294A	1950
3	CTG_HLR	294A	3CD0
4	CTG_IN	294A	1BD0

From the Table 1, we see that the origination point code (OPC) for Bogra MSC is 294A. It is a unique hexadecimal number for a particular SP around the country provided by Bangladesh Telecom Regularity Commission (BTRC). From Table 1, we also see point code for DHK_MSC, CTG_MSC, CTG_HLR and CTG_IN are respectively 1150, 1950, 3CD0 and 1BD0. The Destination Switching Point (DSP) is indexed here serially for those SPs from 1 to 4.

Table 10-2: MTP Link Table Information

Link No.	Link Name	SLC	Link Set No.
1	BOG_MSC to DHK_MSC_1	0	1
2	BOG_MSC to DHK_MSC_2	1	1
3	BOG_MSC to CTG_MSC_1	0	2
4	BOG_MSC to CTG_MSC_2	1	2
5	BOG_MSC to CTG_HLR_1	0	3
6	BOG_MSC to CTG_HLR_2	1	3
7	BOG_MSC to CTG_IN_1	0	4
8	BOG_MSC to CTG_IN_2	1	4

In the Table 2, the signaling links are configured. Link No. is indexed serially in a SP. There are two signaling links configured between BOG_MSC and DHK_MSC. Link No.1 and 2 are configured for BOG_MSC to DHK_MSC for some E1's. Generally, for each E1, timeslot-16 is kept for signaling purpose and timeslot-0 is reserved for synchronization. But, when a group of E1's are configured as a single attribute or an office, then suppose for One STM-1 (63E1), there may be used 3 or 4 signaling links (SL) are practically used. Link Set No. is same for that particular attribute or office. SLC is numbered as 0 and 1 because of uniqueness for a specific Link Set No. So, total number of signaling links in a particular attribute or office should be marked serially as SLC indexing increasing 0, 1, 2……so on.

Table 10-3: MTP Route Table Information

Route Name	DSP Index	Link Set	Priority
DHK_MSC_Direct	1	1	0
CTG_MSC_Direct	2	2	0
CTG_MSC_STP_DHK_MSC	2	1	1
CTG_HLR_ Direct	3	3	0
CTG_HLR_STP_DHK_MSC	3	1	1
CTG_HLR_STP_CTG_MSC	3	2	2
CTG_IN_ Direct	4	4	0
CTG_IN_STP_DHK_MSC	4	1	1
CTG_IN_STP_CTG_MSC	4	2	2

In Table 3, MTP signaling links are configured for routing. For each SP to another SP, there are two types of signaling links, i.e. ~ Direct and STP links are configured.

Now, we will explain about some cases about MTP routes.

Case 1:

From BOG_MSC to DHK_MSC, there is a configuration for only direct link (DHK_MSC_Direct). From Table 1, it is already known that Dhaka MSC DSP index is 1, Signaling link set no is 1 and for direct link the priority is 0. As there is no STP connection, so there is no further configuration.

Case 2:

Now, for connection between BOG_MSC and CTG_MSC, there are two types of signaling links, i.e.~ CTG_MSC_Direct & CTG_MSC_STP_DHK_MSC. From Table 1 and Table 2, it is already known that CTG_MSC DSP index is 2, Signaling link set no is 2 and priority for direct link is 0. For STP_DHK_MSC, Signaling link set no is 1 and priority for STP link is 1.

Case 3:

For BOG_MSC to CTG HLR connection, there are three routes, i.e. ~ one direct and two STP links configured.

When connection with CTG_HLR_Direct, DSP index for CTG_HLR is always 3 (Table 1), Link Set No. is 3 (Table 2) and for direct connection the priority is 0.

For CTG_HLR_STP_DHK_MSC, DSP index is 3, Link Set No. for DHK_MSC is 1 and it is provided second priority 1.

For CTG_HLR_STP_CTG_MSC, DSP index is 3, Link Set No. for CTG_MSC is 2 and it is provided third priority 2.

Case 4:

For BOG_MSC to CTG IN connection, there are three routes, i.e. ~ one direct and two STP links configured.

When connection with CTG_IN_Direct, DSP index for CTG_IN is always 4 (Table 1), Link Set No. is 4 (Table 2) and for direct connection the priority is 0.

For CTG_IN_STP_DHK_MSC, DSP index is 4, Link Set No. for DHK_MSC is 1 and it is provided second priority 1.

For CTG_IN_STP_CTG_MSC, DSP index is 4, Link Set No. for CTG_MSC is 2 and it is provided third priority 2.

These are the explanations of practical implementation of SS7 links configuration for a SP in a core network.

10.3 Summary

In the previous section, the SS7 signaling network is described and explains the practical implementation of MTP links in core network of an operator. Different SP (MSC, HLR, SCP, IN, SMSC) are normally implemented in the same core network. For simplicity, we describe here MSC, HLR and IN for direct and STP connection and how the MTP links are configured in the switch. These case studies explanation helps anyone for understanding SS7 link characteristics such as link configuration and link routing.

Acknowledgement [22]

This network diagram (Figure 10-1) and Data (Table 10-1, 10-2, 10-3) are taken partially from core network department of RankTelecom Limited, the largest PSTN operator in Bangladesh. We would like to thanks all the members of core network department.

Chapter 11

Conclusion and Future Recommendations

11.1 Conclusion

Signaling System 7 (SS7) is a packet-switched data network that forms the backbone of the international telecommunications network. SS7 plays an important role in both wired and wireless networks. It was designed to improve network operation and to provide enhanced services. In this paper we have discussed about the evolution, application, services, advantages, network and protocol architecture, message transfer part, user parts.

In chapter 1 we have studied about the evolution of signaling system 7(SS7). SS7 is a common channel signaling version and the previous version of it is SS6. Common Channel signaling protocols was developed by major telephone companies and the ITU-T since 1975 and was declared as an international standard by ITU-T in its 1980. Different types of signaling and their corresponding advantages and limitations are discussed here.

Chapter 2, 3 and 4 consists of the roles of SS7 in telecommunication networks, the signaling architecture and the various protocols available in SS7. Different types of link and link sets, routes and route sets were the highlights of chapter 3 including the importance of nodes and different types of links.

MTP level 1 relates to the functionalities of physical connections and also defines the medium characteristics which was resolved in chapter 5. MTP level 2 provides the data link layer functionality like Delimitation, Length Indication, Alignment, Error Detection, Error Correction, Flow Control etc. The MTP level 3 performs the route management, alternate routing, link and traffic management. In chapter 6 the SCCP protocol and its architecture are introduced. The connection less and connection oriented services, SCCP message structure and massage types are discussed. Basic call flow and message flow are two important part of ISUP. ISUP services, different types of message flow like IAM, ACM, ANM, REL, and RLC are presented in chapter 7. Message tracing was another very important topic discussed here. TCAP structure, role of TCAP in call controls transaction and sub layers and different messages are discussed in chapter 8. The MAP services like Mobility, Short Message Service (SMS), Call Handling, Supplementary, Operation and Maintenance are briefly discussed in chapter 9.

In chapter 10, practical implementation of SS7 signaling links in the core network of telecommunication sector for an operator is explained.

11.2 Future Recommendations

In this paper we have tried to discuss the protocols as well as their different message types and services of SS7 very briefly. In future it could be possible to optimize different services of SS7 by further studying it and can improve one's ability for the implementation of SS7 in telecommunication networks.

REFERENCES

1. ITU-T Rec. Q.9 (11/88) Vocabulary of Switching and Signaling Terms.
2. Dryburgh, L. and Hewett, J. (2004) *Signaling System No. 7 (SS7/C7): Protocol-Architecture and Services.* CISCO Press.
3. http://en.wikipedia.org/wiki/Channel_Channel_Signaling
4. RFC 2719 - Framework Architecture for Signaling Transport
5. http://en.wikipedia.org/wiki/Channel_Associated_Signaling
6. http://www.phonelines2u.co.uk/marketing-numbers.html
7. www.ss7-training.net
8. http://en.wikipedia.org/wiki/Enhanced_9-1-1
9. www.etisalat.ae ;
10. Harte, L., Dreher, R., Bowler, D. and Beninger, T. (2003) *Signaling System 7 (SS7) Basics*. 3rd Edition. Altos Publishing.
11. http://www.wirelessdevnet.com/channels/sms/features/sms.html
12. http://www.cellular.co.za/technologies/ems/ems.htm
13. http://en.wikipedia.org/wiki/Virtual_private_network
14. Long, J. (2002) *Crackdown on Telemarketers Equals Risk, Opportunity for Telecos. Phone+*
15. http://en.wikipedia.org/wiki/Mobile_Application_Part
16. HUAWEI M800 CDMA Mobile Switching Center, Technical Manual, Vol. Signaling System, Manual Version: T2-030239-20060208-C-6.32 Product Version: V610R003, BOM: 31026239
17. www.javvin.com/protocolSS7SCCP.html
18. http://www.javvin.com/protocolISUP.html
19. www.telecomspace.com/ss7-isup.html
20. http://en.wikipedia.org/wiki/ISDN_User_Part
21. Introduction to Signaling System No. 7, publication number TEC-GEN-003 Rev. B, 2001, SUNRISE Telecom Incorporated.
22. Mowla,M.M., Suman,A.A. and Paul,L.C.(2012) 'Practical Implementation of SS7 Signaling Link in Telecom Core Network', *1st International Conference on Electrical, Computer and Telecommunication Engineering (ICECTE).* RUET, 01-02 December. RUET: Faculty of ECE, RUET, Bangladesh, pp. 42-43.

23. www.iec.org
24. Ronayne, J. P. (1986) *The Digital Network: Introduction to Digital Communications switching* (1st Edition) Indianapolis: Howard W. Sams & Co.

www.ingramcontent.com/pod-product-compliance
Lightning Source LLC
Chambersburg PA
CBHW080916170526
45158CB00008B/2128